WEATHER RADIO

A Complete Guide To Receiving NOAA, VOLMET, WEFAX, Weather Satellites And Other Weather Information Sources

Anthony R. "Tony" Curtis, K3RXK

TIARE PUBLICATIONS

Copyright© 1992 by Tiare Publications

All rights reserved. No part of this book may be reproduced or transmitted in any form or by any means electronic or mechanical, including photocopying, recording, or by information storage or retrieval systems, without permission in writing from the publisher, except by a reviewer who may quote brief passages in a review.

Book design by *Next Wave Graphics*, Caledonia, NY

Published by Tiare Publications, Post Office Box 493, Lake Geneva, Wisconsin, 53147

ISBN 0-936653-35-5
Printed in the United States of America

```
                 Library of Congress Cataloging-in-Publication Data


Curtis, Anthony R., 1940-
    Weather radio   Anthony R. "Tony" Curtis.
       p.   cm.
    "A Complete guide to receiving NOAA, Volmet, weather Fax, weather
satellites and other weather information sources."
     ISBN 0-936653-32-9 (pbk.) : $14.95
     1. Weather reporting, Radio--United States--Handbooks, manuals,
etc.  2. United States.  National Weather Service--Communication
systems--Handbooks, manuals, etc.  3. United States.  National
Oceanic and Atmospheric Administration--Communication systems-
-Handbooks, manuals, etc.  4. Weather forecasting--United States-
-Handbooks, manuals, etc.  5. Television weathercasting--United
States--Handbooks, manual,s etc.   I. Title.
QC875.U7C87  1992
551.6'3'0973--dc20                                           92-3199
                                                                 CIP
```

Table of Contents

Preface	4
Chapter One: Weather Forecasting	**5**
The Telegraph	5
The Weather Bureau	6
Radio Comes To Forecasting	6
Remote Sensing	7
Predicting Weather	8
World Weather Watch	8
Weather Extremes	9
Monitoring Severe Weather By Radio	12
Chapter Two: NOAA Weather Radio	**13**
NOAA Bulletin Boards	13
Chapter Three: Weather Satellites	**14**
Explorer And TIROS Weather Satellites	14
USSR, Nimbus And ESSA Weather Satellites	15
NOAA Weather Satellites	16
The Orbits Of Satellites	17
SMS and GOES Weather Satellites	18
USSR And European Weather Satellites	21
Popular Weather Satellite Frequencies	22
Japan, China, India Weather Satellites	22
Other Earth-Observing Satellites	23
Observation-Satellite Frequencies	24
Weather Satellites Do Search And Rescue	24
How To Receive Weather Satellite Signals	27
WEFAX On TVRO	30
Source List Of Equipment Hardware And Software Manufacturers	31
Chapter Four: Shortwave WEFAX	**33**
Shortwave WEFAX Frequencies	33
WEFAX Transmission Schedule	35
Chapter Five: Hurricane Hunters	**37**
Hurricane Hunter Frequencies	37
Hurricane Disaster Potential	38
VORTEX Voice Message Codes And RECCO Reporting Codes	38
Chapter Six: VOLMET And Aviation Weather	**41**
VOLMET Frequencies, Stations And Times	41
FAA, Military And Coast Guard Weather	42
ATIS And AWOS	43
Chapter Seven: Maritime Weather Services	**44**
Propagation: How Shortwave Radio Signals Skip Around The Globe	44
NAVAREA Warnings	45
Public Coast And AT&T High-Seas Radiotelephone Stations	46
Iceberg Patrols	47
NAVTEX	48
Worldwide Standard Time Stations	48
Russian Fishing Fleets Weather	51
Chapter Eight: Land Mobile Radio And TIS	**52**
Land Mobile Radio Meteorological-Use Frequencies	52
Weather On Traveler's Information Service Stations	53
Chapter Nine: Amateur Radio In Weather Crises	**54**
Amateur Radio Weather-Emergency Net Frequencies	54
SWL Nets For Frequency Updates, W1AW Bulletins, Wind Profilers	87
Chapter Ten: Master List Of Weather Radio Frequencies	**88**
Wind Chill And Heat Index	110
Index	**112**

Preface

If you're like me, every time a major storm threatens you blip through TV channels and run up and down the am-fm radio dial, hoping for a scrap of news about the approaching tempest. Finding remarkably little to satisfy the hunger to know more, sooner, you feel there has to be some way to find out even more current information. But how?

Radio! Not just your local am-fm entertainers, but longwave, shortwave, vhf, uhf, even microwave. *WEATHER RADIO* is a road map to the excitement beyond the weather channel on your TV cable. The idea is to find just about every frequency where there might be a remote possibility of hearing weather reports, data and forecasts. This is a handy look-up guide to where the action can be found in the radio spectrum—how to track storms through radio monitoring and get the depth of advanced weather reporting you need.

After a snap course in the art of weather forecasting, this handy book will introduce you to the mind-expanding worlds of receiving photos from weather satellites, weather maps from shortwave fax stations, aviation and marine weather, hurricane hunter aircraft, Traveler's Information Service, industrial-strength land mobile radio, amateur radio in weather crises, tornado and hurricane watch nets, local highway and emergency rescue crews, even NOAA Weather Radio and hundreds of other weather information sources.

From the Voice of the National Weather Service to strange-sounding acronyms like VOLMET, WEFAX, TIS, NAVTEX, SKYWARN, you'll learn where to tune when foul weather strikes. To stay in touch, radio insiders tune directly to two-way weather communications among those hardy souls out on the front lines fighting the storm. *WEATHER RADIO* reveals how you can monitor that all-important late news long before it airs in local TV station broadcasts or is printed in local newspaper reports.

This reference database details how NOAA's National Weather Service works, how global weather data is collected, and how radios and satellite fax are tools for safety. While others fret, your communications radio will let you look forward to see what may happen in coming hours or days. Whether looking at faxed weather maps or scanning for hurricane hunters, you can get the information you want right from the horse's mouth. You'll even be able to monitor professional news media on your radio by tuning to TV station audio.

Some frequencies change over time, of course, so we can't guarantee you'll find all the info you need at any one place on your radio dial. Just keep on tuning. You'll find what you want on an alternative channel. We've tried to include more than plenty of everything for your listening pleasure, but you never can know 'em all. The author would be grateful to hear of other stations found on alternative frequencies and in out-of-the-way locales across the nation and around the globe. If you have suggestions for future editions of this book, please address the author in care of Tiare Publications, Post Office Box 493, Lake Geneva, Wisconsin 53147 USA.

By the way, two-way point-to-point communications are different from one-way broadcasts intended for dissemination to the general public. Privacy of point-to-point communication is protected by federal law. Listeners are forbidden to reveal or repeat to anyone the content of any two-way radio communication, except amateur radio communications. The prohibition includes reporting anything heard to news media.

In the meantime, keep listening for hurricanes, tornadoes, blizzards and even those sunny days at the beach. A world of exciting armchair adventures is ready to be captured. Get a receiver, put up an antenna, and enjoy. You never know what you'll hear!

—*Anthony R. "Tony" Curtis, K3RXK*

Weather Forecasting

So easy to say, so hard to do: weather forecasting is the prediction of future weather. Meteorology is a science—the study of Earth's atmosphere and variations in temperature and moisture patterns which produce weather conditions. Forecasting future weather is the main mission of the science of meteorology.

Before the telegraph. The word meteorology is from the Greek for astronomical phenomenon—meteoron. Aristotle's Meteorologica in about 340 B.C. covered everything above the ground, but astronomy later was split off into a separate science with meteorology restricted to study of the atmosphere.

For hundreds of human generations, weather forecasting was limited to on-the-spot local observations of temperature, wind and sky conditions and a memory of past climate conditions. Records accumulated only as weather lore. Only in recent generations, has the weather forecaster's horizon been expanded by technology advances.

For nearly 2,000 years, Aristotle's Meteorologica was the standard reference—until Rene DeCartes, Galileo Galilei and other inquiring minds of the 17th century began to go beyond conjecture to instrument observations. The barometer, hygrometer, and thermometer were invented in the century between 1650 and 1750.

Scientific barricades fell like ten pins in the 17th, 18th and 19th centuries:

* Issac Newton devised laws of motion, cooling and refraction,
* Blaise Pascal, Edme Marriotte, Robert Hooke, Edmund Halley, and others calculated how to measure altitudes precisely (hypsometry),
* Robert Boyle studied gases,
* Edmund Halley, John Hadley and Jean Le Rond d'Alembert drew up models of atmospheric circulation,
* thermometers were standardized between 1750 and 1850,
* Benjamin Franklin studied lightning,
* John Dalton measured evaporation and humidity,
* Luke Howard devised a classification system for clouds.

The Telegraph

A new wave of technology through the 1800s encouraged weather observations by individuals as well as public institutions.

* Cup and pressure anemometers were invented.
* Electricity was used to record instrument readings.
* More reliable clocks led to continuous recordings.
* After the telegraph was invented in 1937, forecasters were able to receive weather observations from distant locations within hours. Meteorologists across Europe and America started exchanging weather data via telegraph after a French fleet was damaged by a storm in the 1853-56 Crimean War.

Accumulated weather reports were used to draw "synoptic" weather maps displaying data for one time period over a large area.

Forecasters saw that weather systems move and precipitation comes with low atmospheric pressure, but the physics of atmospheric motions were not understood. Predictions were based on a mixture of empirical rules.

The Weather Bureau

U.S. government meteorology started in 1870 when Congress directed the Army to forecast storms over the Atlantic and Pacific coasts and the Great Lakes. From 1870-1891, it was known as the General Weather Service.

Farmers needed forecasts of weather and long-term climate trends so forecasting was moved to a new civilian Weather Bureau in the Department of Agriculture in 1891. From 1891-1970, it was known as the U.S. Weather Bureau.

Aviators needed frequent observations and short-term forecasts so the Weather Bureau was transferred to the Department of Commerce. In 1965 the Bureau became part of a new Environmental Science Services Administration (ESSA), with climatology separated into a new Environmental Data Service (EDS).

In 1970, the Bureau was renamed the National Weather Service (NWS), a part of the National Oceanic and Atmospheric Administration (NOAA) headquartered in Rockville, Maryland. NOAA was set up within the U.S. Department of Commerce during a reorganization of the federal government in 1970.

The National Weather Service operates the National Meteorological Center at Camp Springs, Maryland, National Hurricane Center at Coral Gables, Florida, and National Severe Storms Forecast Center at Kansas City, Missouri.

NWS has hundreds of meteorological, oceanographic and hydrological stations which release millions of forecasts about weather in the United States and its possessions to the public each year. NWS warns the public about destructive natural events such as floods, tornadoes and hurricanes.

NOAA also conducts geodetic surveys and environmental-information services for the public. It operates the National Ocean Survey and National Marine Fisheries Service.

Radio Comes To Forecasting

Accumulating data over large areas led to more complete descriptions of the atmosphere.

✻Kites, balloons and planes carried instruments through the lowest layer of the atmosphere—the troposphere—to the second lowest layer—the stratosphere—which was discovered just after 1900.

✻Radiosondes are robot instruments carried aloft by balloons, transmitting observations to the ground while ascending through the atmosphere. Development of lightweight, battery-powered radios brought systematic upper-air observations by the end of the 1920s. The clearer radiosonde pictures of the atmosphere revealed features such as the jet steam.

Synoptic meteorology dominated from 1850-1950, with laws of physics replacing miscellaneous empirical rules. Physicists studying thermodynamics after the 1850s wrote formulas describing atmospheric motions and transformations.

Dynamic equations were written by Norwegian scientist Wilhelm Bjerknes in 1904. Fronts and air masses—the polar-front theory of cyclones—came to be understood in the 1920s by Bjerknes and his son Jacob.

The polar-front theory pictures atmospheric circulation systems such as cyclone and anticyclone and the formation of precipitation. It revolutionized weather forecasting by accounting for large-scale movement of air masses.

American meteorologist Jule Charney gave birth to modern meteorology in 1948 when he simplified the dynamic equations. His numerical weather prediction forecasts phenomena by solving the equations which govern the behavior of the atmosphere.

New digital computers accelerated Charney's weather forecasting techniques. Numerical forecasts in 1950 were so good they led to the computerized systems and numerical models which are central to weather forecasting today.

Remote Sensing

Technology for remotely sensing the atmosphere expanded after 1948:

✳By the 1950s, radar reflections were used to find clouds by their suspended water droplets. Radar echoes revealed the internal structure of thunderstorms. Doppler-shift radar added velocity measurement in the 1960s.

✳After 1960, orbiting weather satellites could radio Earth observations from space.

✳Today, meteorology is increasingly computerized and automated, which creates a flood of observations from the mix of old and new instruments. Fast computer processing of Doppler radar information increases warning time for tornadoes and severe weather.

✳NEXRAD—Next Generation Radar—is a digital Doppler radar new in the 1990s. NEXRAD detects motion of the atmosphere toward or away from the radar.

✳Also new in the 1990s, Wind Profiler is a wind-sensing network, which uses the Doppler effect to continuously describe the profile of wind—that is, measure horizontal wind speeds at various heights above the ground.

✳Scatterometer is a new satellite sensor, which measures ocean surface wind speed. It and other new satellite sensors in the 1990s will increase by 100 to 1,000 times the amount of weather information received from space.

✳New automated land, ship and aircraft stations, and robot ocean buoys, will make more-thorough observations in the 1990s. Radios to transmit data, and high-speed processing and display computers, are being developed to handle the torrent of data from these automated systems.

Today, reports from observation stations on land, ships at sea, aircraft, radar, radiosondes and meteorological satellites are transmitted via the World Meteorological Organization's Global Telecommunications System to regional and global meteorology centers for use in numerical weather-forecasting models.

NOAA's Program for Regional Observation and Forecast Systems (PROFS) uses interactive computer systems at Boulder, Colorado, to handle the mass of incoming data.

Numerical models start from data observed at 0000z and 1200z Coordinated Universal Time (UTC). Weather graphs or plots are printed to assist forecasters.

Universal Time. Around the globe, time changes one hour for each 15 degrees of longitude. We are accustomed to thinking of the local time where we are, but there is a single time usable anywhere—Coordinated Universal Time (UTC), formerly Greenwich Mean Time (GMT). UTC is expressed as 24-hour time.

The four-digit 24-hour clock, sometimes called military, government, transportation or official time, designates hours from Midnight to Noon as 0000 to 1200 and from Noon to Midnight as 1200 to 2400. The first two digits are hours, the last two are minutes. That is, the hours 1 p.m. to 11 p.m. are 1300 to 2300z. The small z attached to a time indicates UTC and, often, is spoken as "zulu."

0000z and 2400z are the same time. 0000z is used to indicate the start of a day. 2400z is used to indicate the end of a day. To convert from 12-hour time to 24-hour time, the hours after Noon are added to 12. For example, 3 p.m. is 1500 and 8:30 p.m. is 2030. *(See Time Signal Stations, chapter 7, pages 48-51.)*

Predicting Weather

Modern meteorologists focus on weather patterns—rain, snow, thunderstorms, tornadoes, cyclones, hurricanes, typhoons, monsoons—to understand the origin of specific events. Climatologists study the origins of atmospheric patterns over time.

Accurate forecasting is important when weather events have important consequences. A drought, for instance, might damage crops, dry fire-prone grasslands and forests, lower public water supplies, reduce the flow of rivers, restrict water recreation and travel, let saltwater infiltrate coastal bays and aquifers, raise stress on plant and animal species, bring economic hardship which could shift human populations and force political unrest.

Weather forecasters follow a set of procedures:

✸observation of the atmosphere,

✸analysis of the observations,

✸extrapolation of the future atmosphere assuming weather features will continue to move as they have been moving,

✸prediction of local events.

World Weather Watch

The International Meteorological Organization was founded in 1873 to collect weather information from around the globe using a then-new technology—the telegraph. After World War II, the new United Nations converted the old International Meteorological Organization into a new World Meteorological Organization (WMO) to standardize weather reporting. Today, WMO synchronizes data from weather services in nearly 200 nations.

WMO started planning in 1963 for a World Weather Watch, which came to life in 1968 as a network of surface, upper-atmosphere, and satellite observations. World Weather Watch (WWW) technicians at international centers dotted around the globe study weather patterns to improve forecasting. The centers compute large-scale weather system models to send forecasts through the network. The National Meteorological Center at Suitland, Maryland, and the European Center for Medium-Range Weather Forecasting at Bracknell, Great Britain, are well-known WWW centers.

The Tropical Ocean Global Atmosphere Project links tropical oceans and the atmosphere while the International Satellite Cloud Climatology Program measures the effects of clouds on the atmosphere. World Meteorological Organization sends reports around the world via its Global Telecommunications System (GTS).

The atmosphere is complex and numerical models are not perfect, so predictions are less reliable further into the future. Forecasts beyond three weeks are worthless.

✸Short-range forecasts—nowcasts—cover twelve hours into the future.

✸Daily-range forecasts are for one to two days into the future.

✸Medium-range forecasts cover three to seven days into the future.

✸Extended forecasts combine numerical models with statistical predictions to extend forecasting more than a week ahead.

✸Short-term climate forecasts are one-month and three-month statistical predictions of average forecasts by NWS's Climate Analysis Center.

Weaknesses in forecasting. Forecasting is complicated by motion of the atmosphere—the compressible fluid air—as it rotates around the globe. Water in the atmosphere changes between solid, liquid and gas. Energy from the Sun is the driving force in the atmosphere. The chemistry is changed by human activity—ozone, carbon dioxide, acid rain. Equations extrapolate future behavior of the atmosphere. More exact extrapolation is more expensive as a computer takes more time to solve equations.

Major sites. A mainframe computer at NWS's National Meteorological Center (NMC) at Suitland, Maryland, calculates four forecast models twice a day from data gathered at 0000z and 1200z. Two models cover the entire globe. Extra models are computed during hurricanes and other times as needed. Results are transmitted to NWS offices, private meteorologists, universities, government agencies, international centers via Global Telecommunications System and the general public.

European forecast models are constructed, with more detail and costlier approximations than any other around the globe, at the European Center for Medium-Range Weather Forecasting (ECMWF), Bracknell, England. Calculations are made once a day from 0000z data. Results are sent to member nations and some via Global Telecommunications System.

Lowest Temperatures

Land Mass	Locale	Fahrenheit	Celsius
Africa	Ifrane, Morocco	-11	-24
Antarctica	Vostok	-129	-89
Asia	Oymykon, Russia	-90	-68
Australia	Charlotte Pass, New South Wales	-8	-22
Europe	Ust'Shchugor, Russia	-67	-55
Greenland	Northice	-87	-66
North America	Snag, Yukon, Canada	-81	-63
Oceania	Haleakala Summit, Maui, Hawaii	14	-10
South America	Sarmiento, Argentina	-27	-33

Highest Temperatures

Land Mass	Locale	Fahrenheit	Celsius
Africa	Azizia, Libya	136	58
Antarctica	Esperanza, Palmer Peninsula	58	14
Asia	Tirat Tsvi, Israel	129	54
Australia	Cloncurry, Queensland	128	53
Europe	Seville, Spain	122	50
North America	Death Valley, California	134	57
Oceania	Tuguegarao, Philippines	108	42
South America	Rivadavia, Argentina	120	49

Australia, Canada, China, Great Britain and Russia also construct regional and global numerical forecast models daily. Other countries only use forecasts available from Global Telecommunications System.

Local predicting. When a local forecaster wants to predict the low temperature for a given night or some other specific variable, lots of data is available—both locally observed and computer modeled.

Data, however, is not a prediction. The forecaster also must look to models of past situations, average conditions and local variations.

Taking local weather history into account by using so-called statistical regression equations with coefficients which vary by location and season, National Meteorological Center forecasts include statistical predictions of precipitation, wind, temperature and other variables for 286 U.S. weather stations.

Weather Extremes

Weather is the state of the atmosphere—temperature, precipitation, humidity, cloudiness, visibility, pressure, winds—varying over periods from minutes to months. Climate, on the other hand, is the weather of a particular region.

The Sun drives Earth's weather changes from day to night (diurnal cycle) and year to year (annual cycle). In the course of a year, Earth's motion around the Sun brings daily, monthly, and seasonal changes.

The diurnal cycle varies by climate and season. Trade winds may change Pacific island day-to-night temperatures by 11 degrees Fahrenheit, while someplace in the center of the North American continent may change by 23 degrees in summer and 16 degrees in winter.

The greatest temperature change in 24 hours in the U.S. occurred when the temperature in Billings, Montana, fell 101 degrees, from 46 degrees to –55 degrees Fahrenheit.

Weather also differs with distance or intervening bodies of water, as when neighboring snowfall amounts differ. For instance, Grant Park, California, on a mountain at 6,580 feet above sea level, is 20 degrees cooler in July and has four times the precipitation of nearby Fresno, California, which is only 330 feet above sea level.

Milwaukee, Wisconsin, on the western shore of Lake Michigan, has fewer cloudy days, fewer thunderstorms, less snowfall and a different prevailing wind direction, than Grand Haven, Michigan, on the eastern shore of the lake.

Conditions which vary excessively are said to be extreme. Highest and lowest values are absolute extremes. The highest temperature ever recorded on Earth was 136 degrees Fahrenheit in the Libyan desert. The lowest was –129 degrees in Antarctica. The most rainfall in one year was 460 inches in Hawaii. The least rainfall was 0.03 inches in the desert in Chile.

Some places have wide-ranging extremes. For example, Fairbanks, Alaska, recorded a high of 99 and a low of –66, a difference of 165 degrees. But, you don't have to travel in Alaska to find such a wide range of extremes. Boise, Idaho, recorded a high of 114 and a low of –45, a range of 159 degrees.

U.S. residents see many of the world's climates—tropical, desert, continental, Arctic. Florida is subtropical with high temperatures, heavy rainfall, and numerous thunderstorms.

Hawaii's temperatures are like Florida, but its tropical climate is dominated by trade winds, resulting in little day-night or month-to-month temperature change, moderate precipitation, and almost no storms.

The San Francisco Bay area's maritime climate is affected by cold ocean currents, resulting in little month-to-month temperature difference, a winter rainy season and few storms. Arizona has a desert climate with hot summers and very little rain.

At the center of the continent, Missouri has a humid continental climate, like the eastern U.S., with great seasonal temperature differences, moderate precipitation every month and many thunderstorms during warm weather. Alaska has an Arctic continental climate with a very cold winter and very little precipitation.

Weather conditions over any land area anywhere on Earth are determined by several factors:

✱solar energy received,
✱nearness to mountains,
✱elevation,
✱proximity to large bodies of water,
✱temperature differences between land and nearby water,
✱number of big storm systems caused by differences in air-mass,
✱distribution of air pressure over land and the nearest ocean,
✱varying wind and air-mass patterns.

Most Average Annual Precipitation

Land Mass	locale	inches	cm
Africa	Debundscha, Cameroon	405	1028
Asia	Cherrapunji, India	450	1143
Australia	Tully, Queensland	179	455
Europe	Crkvica, Yugoslavia	183	465
North America	Henderson Lake, B.C., Canada	262	665
Oceania	Mt. Waialeale, Kauai, Hawaii	460	1168
South America	Quibdo, Colombia	354	899

Least Average Annual Precipitation

Land Mass	Locale	inches	cm
Africa	Wadi Halfa, Sudan	0.1	.25
Antarctica	South Pole Station (solid snow, liquid content unknown)	0.8	2.03
Asia	Aden, Saudi Arabia	1.8	4.57
Australia	Mulka, South Australia	4.05	10.29
Europe	Astrakhan, Russia	6.4	16.26
North America	Batagues, Mexico	1.2	3.05
Oceania	Puako, Hawaii	8.93	22.68
South America	Arica, Chile	0.03	.08

North America has a greater range of latitude than of longitude, which creates a lot of differential heating, fostering differences in air masses. Air masses include continental Arctic and polar air from over cold land, maritime polar air from over cold oceans, and warm Gulf and Atlantic ocean air.

Pacific Ocean air affects weather in western mountains, which affects the interaction of cold and warm air in the east. The way the Gulf of Mexico is carved into the southern states affects air masses, pressure centers and storm systems over the eastern half of the U.S. Moist warm air from the south often collides over the central U.S. with dry cold air from the north.

Movements of these air masses across the North American continent change the weather, bringing sunshine, cloudiness, rain, snow, thunderstorms, ice storms, sleet, hail, tornadoes, floods, droughts and other forms of mild and severe weather.

Thunderstorms carry about half of America's total precipitation—40 percent in the East, 50 percent in the Midwest, 65 percent in the Great Plains and 80 percent in drier mountain climates.

Monitoring Severe Weather By Radio

Of greatest interest to radio-monitor hobbyists are unusual and severe weather events. NOAA weather forecasters pay close attention to their predictions during times of severe conditions such as blizzards, hurricanes and tornadoes. The National Weather Service commits special resources to forecasting such events.

✳Thunderstorms, tornadoes, hail, lightning and downbursts are monitored by the Severe Storm Forecast Center (SSFC) at Kansas City, Missouri, which may issue a convective outlook a day ahead of thunderstorm activity. A convective outlook maps the area of expected activity. Detailed forecasts can be made only one to three hours in advance. When a severe storm is sighted by local National Weather Service offices, government public service agencies or private SkyWarn observers, the SSFC warns local NWS offices, government authorities and news media about places in the path of the storm.

✳Hurricanes in the Atlantic Ocean, Caribbean Sea, Gulf of Mexico and eastern Pacific Ocean are tracked by the National Hurricane Center (NHC) at Coral Gables, Florida. Complementing the orbiting weather satellites, hurricane hunter airplanes fly through the big storms to collect data. National Hurricane Center forecasters use the extra data from hurricane hunters to refine their numerical maps. A big concern of NWS forecasters is the widespread growth of population centers along the Atlantic and Gulf coasts, which would like more than 24 hours warning for evacuation before landfall. Sometimes, forecasting 24 hours in advance is not possible.

✳Blizzards are processed through everyday forecast channels. Local National Weather Service offices issue advisories when needed.

✳Along with monitoring local news media broadcasts, the easiest way for radio hobbyists to track thunderstorms, tornadoes, hurricanes and blizzards is tuning in NOAA weather broadcast stations near 162 MHz *(see chapter 2)*, local highway and emergency rescue crews two-way communications on VHF and UHF public service frequencies *(chapter 10)*, and amateur radio SkyWarn operators sending reports to the National Weather Service via regional VHF and UHF repeaters *(chapter 9)*. In addition, NOAA hurricane hunter aircraft *(chapter 5)* and amateur radio hurricane information nets *(chapter 9)* can be monitored on shortwave frequencies. Satellite weather maps can be received and printed or displayed on computer screen *(chapters 3 and 4)*. Lightning can be monitored directly with so-called lightning detector receivers which react to the crackle of static bursting from nearby lightning strikes. Aviation and marine weather reports and forecasts can be heard *(chapters 6 and 7)*. Local industry activities reflect weather conditions *(chapter 8)*. For a combined master list of weather radio frequencies, see chapter 10.

NOAA Weather Radio

The U.S. National Oceanic and Atmospheric Administration operates hundreds of NOAA Weather Radio stations, known as the "Voice of the National Weather Service."

They broadcast regional forecasts and weather data continuously, using narrow-band fm (nbfm) transmissions on a "weather band" of seven vhf high-band frequencies between 162.400 and 162.550 MHz. More than 90 percent of the U.S. is within range of at least one Weather Radio station. A handful of the stations are operated by private weather-hobbyist groups. Some two dozen stations across southern Canada are operated by the Canadian government on the same band of frequencies as U.S. stations.

NOAA stations usually can be heard up to 40 miles away. Range depends on height of transmitting antenna, terrain between transmitter and receiver, and height of receiving antenna. The higher, the better. In fringe reception areas, outside antennas work better. A few localities have experimental, low-power Weather Radio repeater stations.

NOAA Weather Radio forecasts are tailored for local audiences. For example, stations along sea coasts and Great Lakes add boating, fishing and marine weather forecasts. Stations in farming areas add agricultural weather data.

Taped reports are repeated by a Weather Radio station every five to six minutes and revised every one to three hours, 24 hours a day, seven days a week. A few stations operate less than 24 hours a day, but increase schedules in severe weather.

When severe weather threatens, forecasters interrupt broadcasts to air warnings. They transmit alerting tones to activate special remote-control receivers in listeners' homes and offices. Many specialized weather-band receivers are available. Marine two-way radios and some CB and amateur radio transceivers offer push-button weather band selection.

In 1975, a White House order designated NOAA Weather Radio as the official government-operated broadcaster for direct warnings to private homes in natural disasters or nuclear attack. NOAA broadcasts would supplement commercial radio and TV and local sirens. The list shows frequency and U.S., maritime and Canadian channel designations.

MHz	U.S.	Maritime	Canada
161.650	F-8	Canada-WX 8	WX 8
161.775	F-9	Canada-WX 9	WX 9
162.400	F-2	USA-WX 2	WX 2
162.425	F-4	USA-WX 4	WX 4
162.450	F-5	USA-WX 5	WX 5
162.475	F-3	USA-WX 3	WX 3
162.500	F-6	USA-WX 6	WX 6
162.525	F-7	USA-WX 7	WX 7
162.550	F-1	USA-WX 1	WX 1

NOAA Bulletin Boards

Some computer bulletin board systems (bbs) operated by National Weather Service:

✷The 24-hour East Coast Marine Users bbs for commercial fishermen and others using Mid Atlantic coastal waters. It's open to the public and free of charge. Pre-register at (301) 899-3296. The bbs, at (301) 454-8700, has marine weather for coastal waterways, sounds, bays, coastal and offshore waters, tropical storm advisories, tidal information and priority weather news. Weather for other regions is on other bbs's.

✷NWS's Climate Analysis Center offers a Climate Assessment bbs with historical climate info; daily, weekly, monthly heating degree days; weekly climate bulletins. Pre-register at (301) 763-8071.

✷The NOAA bbs at Boulder, Colorado, for propagation and other information, is at (303) 497-5000.

Weather Satellites

Weather satellites help forecasters see storm systems, weather fronts, jet streams, upper-level troughs and ridges, fog, sea ice, snow cover, upper-level wind direction and speeds and cloud formations. Coastal and island weather stations use satellite data to track tropical storms, hurricanes and typhoons. Visible-light and infrared satellite photos show remote desert, ocean and polar areas where weather would be unknown. The satellites even record ocean surface temperatures for fishing and shipping.

Television news programs often show pictures of cloud patterns snapped from overhead only an hour or so before air time. A weatherman points out differences between places with clear skies and swirling storm clouds. A series of time lapse photos made over a 24-hour period show movement of a storm.

As familiar as they are today, such weather satellite photos were not possible before 1960. That's when Harry Wexler, director of research for the U.S. government Weather Bureau, turned his dream into reality. Wexler had been a leading proponent of weather satellites. He had imagined that cameras in space satellites, orbiting above the clouds, would see thick bands of clouds swarming along weather fronts, popcorn fair weather clouds, and enormous, revolving hurricanes over the oceans.

Explorer. The earliest, experimental TV pictures of Earth's cloud cover were sent down by a satellite called Explorer 6 after it was launched in August 1959.

TIROS. Eight months later, the prototype weather satellite Television and Infra-Red Observation Satellite (TIROS-1) was launched to a 600-mi.-high orbit April 1, 1960.

Wexler was proven correct within hours of the TIROS-1 launch when government scientists showed U.S. President Dwight D. Eisenhower photos of clouds from space. The first weather satellite picture was of clouds over the Gulf of St. Lawrence. The 270-lb. TIROS-1 worked only 78 days, but sent down 22,952 photos.

TIROS became a series of ten experimental weather satellites, launched by NASA between 1960 and 1965 to test television cameras as well as Sun-angle and horizon scanners for meteorology. TIROS-2 was launched in November 1960. TIROS-3 in July 1961. The last in the series was TIROS-10 launched in 1965. The satellites orbited above the equator, carrying two TV cameras each, one with wide-angle lens and one with narrow-angle lens for more detail. Each had an horizon seeker, infrared sensors, tape recorder for storing images, and three radios for transmitting pictures and data to Earth. Solar cells converted sunlight to electricity stored in nickel-cadmium (NiCad) batteries.

TIROS	Launch	Days Of Life	TV Pictures
TIROS 1	1960 Apr 1	89	22,952
TIROS 2	1960 Nov 23	376	36,156
TIROS 3	1961 Jul 12	230	35,033
TIROS 4	1962 Feb 8	161	32,593
TIROS 5	1962 Jun 19	321	58,226
TIROS 6	1962 Sep 18	389	68,557
TIROS 7	1963 Jun 19	1809	125,331
TIROS 8	1963 Dec 21	1287	102,463
TIROS 9	1965 Jan 22	1238	88,892
TIROS 10	1965 Jul 2	730	78,874
Advanced Experimental TIROS			
TIROS M	1970 Jan 23		
TIROS N	1978 Oct 13		

USSR. The Soviet Union launched its first weather satellite, Cosmos 4, in 1962. Cosmos 4 also was the USSR's first photo-reconnaissance spy satellite.

Nimbus. Nimbus is a meteorologist's name for rain clouds. It also was America's second generation of weather satellites. Seven experimental Nimbus satellites were sent to polar orbit between 1964-1978 to take pictures at night as well as during daylight hours.

Nimbus-1, launched in 1964, was the first polar-orbiting U.S. weather satellite. The Nimbus series had better still-photo cameras, television cameras for mapping clouds, and infrared radiometers for night photography. Nimbus satellites flew at 600-mile-high altitudes. An automatic transmission system sent high-resolution TV pictures of cloud cover in visible and infrared light immediately after they were taken. Anyone beneath a Nimbus satellite with a proper radio and fax machine could receive weather photos.

High-resolution infrared radiometers, seeing differences between cloud and surface temperatures, helped meteorologists map night-time clouds. Spectrometers and radiometers measured atmospheric temperature, ozone, water-vapor and solar radiation.

The 10-ft.-tall, butterfly-shaped Nimbus satellites were built by General Electric. Their weights ranged from 830 lbs. to 2,176 lbs. A five-ft. aluminum ring, attached to an upper compartment by a magnesium truss, housed sensors and electronic components and three TV cameras which delivered finer detail than TIROS cameras. An attitude-control system in the upper compartment pointed the cameras toward Earth. Outside were two solar panels sopping up sunlight to produce 550 watts of electrical power for Nimbus.

Some 1,440 pictures of a cross-section of Earth's surface 1,500 miles east-west and 500 miles north-south were taped each day for playback over ground tracking stations.

Nimbus	Launch	Results
Nimbus 1	1964 Aug 28	Transmitted 27,000 cloud-cover photos, until shut off Sept. 23, 1964.
Nimbus 2	1966 May 15	Transmitted infrared and TV cloud-cover photos.
Nimbus B	1968 May 18	Rocket guidance failed, destroyed enroute to orbit.
Nimbus 3	1969 Apr 14	Sent infrared, ultraviolet, television, geodetic data; first U.S. satellite to measure atmosphere day and night temperatures globally at various altitudes.
Nimbus 4	1970 Apr 8	Like Nimbus 3 with new & improved instruments.
Nimbus 5	1972 Dec 11	Like Nimbus 4 with new & improved instruments.
Nimbus 6	1975 Jun 12	Like Nimbus 5 with new & improved instruments.
Nimbus 7	1978 Oct 24	First satellite built to monitor Earth's atmosphere for natural and artificial pollutants.

ESSA. In 1965, the U.S. Weather Bureau became part of a new Environmental Science Services Administration (ESSA) within the Department of Commerce and the next series of nine weather satellites launched from 1966-1969 came to be known as ESSA. The nine satellites formed the so-called TIROS Operational System (TOS).

The ESSA satellites, flying at altitudes around 900 miles above Earth, had the advanced cameras tested early in Nimbus satellites. ESSA cloud-cover photos were recorded in space and transmitted to Earth automatically. Anyone with the simple required equipment could receive them. Other nations installed gear and, today, more than 120 countries get their weather pictures from various U.S. satellites.

ESSA	Launch	ESSA	Launch
ESSA 1	1966 Feb 3	ESSA 6	1967 Nov 10
ESSA 2	1966 Feb 28	ESSA 7	1968 Aug 16
ESSA 3	1966 Oct 2	ESSA 8	1968 Dec 15
ESSA 4	1967 Jan 26	ESSA 9	1969 Feb 26
ESSA 5	1967 Apr 20		

ITOS. The ESSA series of weather satellites was followed by the ITOS series, patterned after TIROS. ITOS stood for Improved TIROS Operational Satellite. The first to orbit was TIROS-M, launched January 23, 1970. In orbit, it became known as ITOS-1.

NOAA. The polar-orbiting satellites which followed ITOS-1 into orbit from 1970-1976 were named NOAA, for ESSA's successor—the National Oceanic and Atmospheric Administration. NOAA-1 was launched December 11, 1970.

NOAA-1 to NOAA-5 were like ESSA satellites. They were launched into 900-mile orbits to see the entire Earth each day. Night cloud photos were made with infrared sensors. Radiometers measured vertical temperature profiles of the atmosphere. Each satellite covered the globe every 12 hours and could relay photos and radiometer data automatically or store them for playback later.

Third Generation. The third generation of U.S. polar-orbiting weather satellites began operation with the launch of TIROS-N in 1978. TIROS-N was a research prototype for the follow-on operational series of satellites which were called NOAA-A to NOAA-M before launch and numbered NOAA-6 and above after launch.

TIROS-N And The Greenhouse Effect

The TIROS-N series of weather satellites has been flying a near-circular Sun-synchronous orbit 530 miles above Earth since 1978, sending down visible-light and infrared-light images with half-mile resolution. A TIROS-N records the temperature profile of the atmosphere from the surface to 20 miles altitude with an accuracy of 2 to 4 degrees Fahrenheit. More than 1,000 ground stations in 80 nations receive pictures.

Twelve years after its launch, TRIOS-N cast doubt in 1990 on the existence of a so-called greenhouse effect around planet Earth. Scientists at NASA's Marshall Space Flight Center and University of Alabama, Huntsville, said analysis of 12 years of data showed no conclusive evidence of global warming from a greenhouse effect.

Beginning with NOAA-6 in 1979, the NOAA name and numbering sequence was continued. NOAA-6 to NOAA-12 are modernized TIROS satellites, part of a cooperative program between Canada, France, Great Britain, NASA and NOAA.

NOAA satellites are equipped with sophisticated cameras and radiometers as well as infrared, stratospheric, and microwave sounders. In addition, they provide readings of sea-surface temperatures, identify snow cover and ice at sea, and measure particle densities in the upper atmosphere in order to predict the onset of solar disturbances.

✷NOAA-A was launched in 1979, renamed NOAA-6, and deactivated in 1987.

✷NOAA-B was launched in 1980, but failed to reach its proper orbit.

✷NOAA-C was launched in 1981, renamed NOAA-7, and deactivated in 1986 after electrical power failed.

✷NOAA-D was dropped out of sequence in favor of NOAA-E, which was a longer spacecraft and could hold more equipment, including a search and rescue radio receiver. NOAA-E was launched in 1983, and renamed NOAA-8 in space. Its clock and electrical power system broke down and it stopped working in 1985.

✷NOAA-F was launched in 1984, and renamed NOAA-9 in space. In 1992, it was on standby in orbit with some data still being processed.

✷NOAA-G was launched in 1986, and designated NOAA-10 in space. It was working well in 1992, except for its Earth Radiation Budget Experiment (ERBE) scanner. The scanner would stick or hang-up sometimes. *(Continued on page 18)*

The Orbits Of Satellites

The U.S. and other nations have dozens of weather satellites orbiting at low altitudes of 400-1,000 miles and high altitudes of 22,300 miles above Earth. Many low-altitude satellites fly east-west in low equatorial orbits—circling the globe above the equator. Others travel a different path, north-south, crossing Earth's poles several times a day. Satellites in polar orbits look down on the entire surface of the Earth every day. As a satellite loops around the globe, Earth seems to rotate underneath.

Observation of the ground is improved if the surface always is illuminated at about the same Sun angle when viewed from a satellite. Most weather satellites are in Sun-synchronous polar orbits—including Nimbus, TIROS, NOAA and Meteor. The first polar-orbiting U.S. weather satellite was Nimbus 1, launched in 1964.

The more distant region of space girdling the globe 22,300 miles above the equator is referred to as the Clarke Belt, after science-fiction writer Arthur C. Clarke who first proposed placing satellites in stationary orbits in 1945. In the Clarke Belt, a satellite travels around Earth at the same speed as the planet turns and appears to hang stationary or stand still in the sky over one spot on the ground. Many weather satellites occupy positions in the Clarke Belt. Geostationary Operational Environmental Satellites (GOES) weather satellites are in stationary orbits, also referred to as synchronous and geosynchronous orbits. Stationary satellites are ideal for monitoring continent-wide weather patterns and environmental conditions.

A satellite can be fired from Earth into a special, highly-elliptical, equatorial orbit known as a geostationary transfer orbit (GTO) where the satellite swings out as far as 22,300 miles and back in to an altitude of 100 miles above Earth. At an assigned time and place along the transfer orbit, a "kick motor" rocket attached to the satellite pushes it on out to a circular orbit at the stationary altitude of 22,300 miles.

Rockets used to launch satellites can be fired in any direction from a launch pad, but human neighbors dictate the direction of launch. Rockets from NASA's Cape Canaveral, Florida, pads mostly are launched east over the Atlantic, but sometimes northeast along the Atlantic Coast. From Vandenberg Air Force Base, California, rockets usually are fired west and north over the Pacific Ocean into polar orbits.

Rockets launched from Baikonur Cosmodrome in Kazakhstan carry satellites to equatorial orbits. Rockets from Russia's Northern Cosmodrome at Plesetsk carry satellites to polar orbits. Rockets from the European Space Agency's Kourou, French Guiana, site near the equator go east or north over the Atlantic to equatorial orbits.

Satellites don't have to be launched from the ground to reach polar orbit. The U.S. fires satellites to polar orbit on a Pegasus rocket carried aloft by a B-52 bomber. The winged rocket was strapped under the airplane's wing, ferried seven miles above Earth and dropped. Pegasus ignited and blasted on up to a 350-mi.-high polar orbit.

The period of a satellite is the time it takes to complete one revolution around planet Earth. The time is shorter for satellites at lower altitudes. Circular-orbit periods:

Period	Altitude	Velocity	
1.5 hours	171 miles	17,311 mph	4.81 mi/sec
3.0 hours	2,597 miles	13,732 mph	3.82 mi/sec
6.0 hours	6,450 miles	10,893 mph	3.03 mi/sec
12.0 hours	12,564 miles	8,656 mph	2.40 mi/sec
24.0 hours	22,270 miles	6,866 mph	1.91 mi/sec

✳NOAA-H was launched in 1988, and renamed NOAA-11 in space. Its instruments were working well in 1992, however its attitude control system had lost two of four gyros and was being controlled by special software.

✳NOAA-D was launched in 1991. In space, it was renamed NOAA-12. Like the others since NOAA-6, NOAA-12 is a TIROS-N class of satellite built by General Electric to provide day and night global environmental data. NOAA-12 has five primary instruments, including the Advanced Very High Resolution Radiometer. The satellite was placed in a 522-mile-high polar orbit inclined 98.7 degrees to the equator.

NOAA	Launch	NOAA	Launch
NOAA 1	1970 Dec 11	NOAA 9	1984 Dec 12
NOAA 2	1972 Oct 15	NOAA 10	1986 Sep 17
NOAA 3	1973 Nov 6	NOAA 11	1988 Sep 24
NOAA 4	1974 Nov 15	NOAA 12	1991 May 24
NOAA 5	1976 Jul 29	NOAA I	scheduled for 1993
TIROS N	1978 Oct 13	NOAA J	scheduled for 1995
NOAA 6	1979 Jun 27	NOAA K	scheduled for 1997
NOAA 7	1981 Jun 23	NOAA L	scheduled for 1999
NOAA 8	1983 Mar 28	NOAA M	scheduled for 2001

Stationary weather satellites. At 22,300 miles above Earth, a satellite would be at a much higher altitude than the polar-orbiting TIROS and Nimbus. In fact, at that great distance, a satellite would seem to hang stationary above Earth's surface.

The U.S. launched the first stationary communications satellite, Syncom, in 1963. After 1970, meteorologists wanted pictures shot more frequently than once or twice a day. They needed a satellite parked overhead, continuously taking pictures from stationary orbit.

NASA built two experimental Synchronous Meteorological Satellites (SMS) to test weather satellites in stationary or geosynchronous orbits. An SMS was an 11-ft.-tall, 6-ft.-diameter, aluminum cylinder weighing 535 lbs. A spin-scan radiometer in the satellite stared down continuously, day and night, to record visible light pictures with one-half-mile resolution and infrared images with five-mile resolution.

SMS-1 was launched in 1974 to a spot 22,591 miles above the equator at 45 degrees west longitude. The satellite later was moved to 75 degrees west longitude. SMS-2 was launched in 1975 to a point above the equator at 135 degrees west longitude.

The two observed the Western Hemisphere continuously, transmitting cloud pictures every 30 minutes. SMS led to the Geostationary Operational Environmental Satellite (GOES) series of stationary weather satellites. GOES-1 was launched in 1975, eight months after SMS-2.

SMS	Launch
SMS-1	1974 May 17
SMS-2	1975 Feb 6

GOES. Meteorologists wanted pictures more frequently than once a day. They asked for a satellite parked overhead to take pictures continuously. The result was Geostationary Operational Environmental Satellite (GOES) built for NOAA and launched by NASA.

The first were launched in the 1970s to stationary orbit 22,300 miles above Earth: GOES-1 in 1975, GOES-2 in 1977, and GOES-3 in 1978.

An improved series of four were launched in the 1980s: GOES-4 in 1980, GOES-5 in 1981, GOES-6 in 1983 and GOES-7 in 1987.

The future GOES-I, GOES-J, GOES-K, GOES-L and GOES-M are to be lobbed to space every two-years through the 1990s, to be renamed GOES-8 to GOES-12 in orbit.

NOAA's National Environmental Satellite, Data, and Information Service (NESDIS) operates and maintains the satellites. GOES ground stations are NOAA's Satellite Operations and Control Center (SOCC) at Suitland, Maryland, and the Command and Data Acquisition station (CDA) at Wallops Island, Virginia. Raw weather data is transmitted from a GOES satellite to the CDA station, which processes the information and retransmits it through the spacecraft to various data users, including the National Weather Service.

Built by Hughes Aircraft Company, GOES are 12-ft.-long, 7-ft.-diameter, 1,382-lb. cylinders containing spin-scan radiometers or vertical atmospheric sounders (VAS) for simultaneous photos in visible light and infrared light. They produce three-dimensional pictures. GOES are powered by solar cells generating 320 watts of electrical power.

The satellites have 16-inch telescopes looking down at visible and infrared light, day and night, to observe clouds, cloud heights, winds and vertical temperature profiles.

GOES transmissions are received by manned and unmanned stations on the ground. Forecasters interpret the data to predict weather, ocean currents and river levels. Cloud pictures seen most often by the U.S. public on television today are from GOES satellites.

While NOAA satellites are lower and see more detail over a smaller area, GOES satellites see an entire hemisphere. GOES can photograph one third of the entire planet every 30 minutes, while NOAA satellites in lower polar orbits might pass over a developing hurricane, for example, only once a day.

GOES shortage. There are supposed to be two GOES satellites watching North America—collecting and transmitting visible and infrared images every half hour from offshore over the Atlantic and Pacific oceans.

GOES-5, launched in 1981, monitored the eastern U.S., while GOES-6, launched in 1983, observed the west. After GOES-5 failed in 1984, NOAA moved GOES-6 to 108 degrees west longitude to monitor the entire country. During hurricane seasons, GOES-6 was moved to 98 degrees west longitude to increase coverage of the Caribbean. A replacement for GOES-5 was destroyed when its rocket failed during launch from Cape Canaveral in 1986. The last GOES on the shelf was launched in 1987 to become GOES-7.

Unfortunately, GOES-6 failed in 1989. Since then, NOAA forecasters have been relying on the single remaining satellite, GOES-7. It is moved over the center of the U.S. in winters to watch Pacific storms affecting Alaska, Hawaii and the West Coast. Ground controllers let it slip in summers toward the East Coast for hurricane season. It takes about four months for the satellite to drift west from the Atlantic coast to a mid-continent position.

GOES-7 was working well, but running low on fuel in 1992. NASA said conservative use might leave enough fuel to keep GOES-7 in place until 1995. Fuel is burned in small thruster rockets fired to keep the satellite at its assigned position in orbit. After GOES-7 runs out of fuel, it will drift away from its position over the equator.

The U.S. declared a weather satellite emergency in 1991. Lease of European and Japanese weather satellites was considered. European Space Agency moved its Meteosat-3 weather satellite farther west over the Atlantic Ocean to allow the U.S. National Weather Service to move GOES-7 west for more complete weather coverage of the U.S.

If the last GOES were to run out of fuel before the next GOES is launched, the U.S. would have to rely on pictures from the low-flying, polar-orbiting NOAA weather satellites, as well as European, Japanese and U.S. military meteorological satellites.

Global stationary weather satellite coverage includes a Japanese satellite over the western Pacific, Russian satellite over the Indian Ocean, European satellite over the eastern Atlantic, as well as the American GOES over the western Atlantic and eastern Pacific.

GOES-NEXT. The current GOES series is to be replaced by an improved flotilla of satellites known as GOES-NEXT. A replacement GOES-NEXT satellite was to have been launched in 1989, but remained on the ground through 1992 with technical problems. Mirror flaws, wiring problems and infrared detector defects had been found.

Computer models and hardware tests found temperature extremes—such as exposure to raw sunlight in space—could warp the surface of the satellite's mirror. Five GOES-NEXT satellites are being built with mirrors to reflect images to various sensors. Engineers coated the mirrors to reduce susceptibility to temperature extremes.

The first of the new GOES series may reach orbit in 1993. Development of hardware for GOES I—including the satellite's primary imaging and sounding instruments—is on schedule for a December 1993 launch, according to NASA's engineering team at Goddard Space Flight Center, Greenbelt, Maryland.

GOES	Launch	GOES	Launch
GOES 1	1975 Oct 16	GOES 7	1987 Feb 26
GOES 2	1977 Jun 16	GOES I	scheduled for 1993
GOES 3	1978 Jun 16	GOES J	scheduled for 1995
GOES 4	1980 Sep 9	GOES K	scheduled for 1997
GOES 5	1981 May 22	GOES L	scheduled for 1999
GOES 6	1983 Apr 28	GOES M	scheduled for 2001

Today's NOAA, Nimbus and GOES satellites are essential to the U.S. National Operational Meteorological System. About one-third of the WEFAX pictures intended for the U.S. come from one of the polar-orbiting NOAA satellites; the rest from GOES.

Weather satellite photos have been useful in emergencies. In 1985, the National Weather Service at Redwood City, California, sent GOES 6 satellite photos to help emergency workers locate Big Sur wildfires.

Military satellites. Many U.S. military weather satellites have been developed secretly and launched over the years after 1960. For instance, a rebuilt Atlas ICBM ferried an 1,815-lb. Defense Meteorological Satellite from Vandenberg Air Force Base, California, to a polar orbit in 1988. It was dropped off in an orbit 527 miles above Earth, replacing one launched from Vandenberg in 1983. The Air Force previously had launched a Defense Meteorological Satellite in 1987 from Vandenberg.

Two such military weather satellites usually are in orbit at a time, circling the globe in a north-south flight path every 12 hours. Tracking storms, they scan 1,600-mi.-wide swaths of Earth's surface, feeding cloud cover and temperature information to the military services. NOAA sometimes also uses the information for civilian weather forecasts.

Latest technology. Meteorologists use satellite data—and data transmitted 24 hours a day from more than a thousand manned and unmanned weather stations—to gauge ozone levels, measure water vapor, survey pollution levels, plot storms, follow jet streams, examine fronts, watch fog, estimate snow and ice cover, monitor river levels and detect forest fires.

The latest American weather satellites record atmosphere and ocean temperatures at various altitudes and depths, gauge rainfall for forecasting of droughts and harvests, survey chlorophyll content for crop health studies, spot forest fires, map ocean currents, see volcanic eruptions, and chart ice in shipping lanes.

In fact, they monitor the total global environment. Some also have search-and-rescue capability. They carry radio equipment listening for faint emergency signals from ships in trouble at sea and downed aircraft.

NASA recently devised a satellite thermometer 100 times more sensitive than earlier equipment. The new infrared radiometer measures ocean surface temperatures from space.

It can tell the difference between the surface of the ocean and the atmosphere immediately above it—a major breakthrough. It had been difficult to draw accurate climate maps before, because sensors couldn't make the distinction between surface and atmosphere. The radiometer gauges sea temperatures by receiving and analyzing infrared light in sunlight reflected naturally from the ocean surface.

Meteorologists say they need the better temperature and wind maps from the radiometer to draw more-detailed pictures showing the interaction of worldwide weather patterns. For example, they hope to learn how storms in the tropics spread to other latitudes. Eventually, they want to understand powerful weather such as El Nino, the unusually-warm water in the eastern Pacific Ocean which can damage fishing industries.

The new infrared radiometer is passive—it only receives light in the infrared part of the energy spectrum. Weather radar, on the other hand, is an active device because it both transmits and receives radio signals to monitor weather patterns.

USSR's Meteor. The first USSR weather satellite was Cosmos 4 launched in 1962. Today's Russian weather satellites also are known as Meteor. Many have been launched to polar orbit in three series. The twentieth Meteor-2 weather satellite was launched in 1990. The fourth and fifth Meteor-3 weather satellites were launched in 1991.

Sometimes There Are Problems...

Private persons who enjoy eavesdropping on foreign satellite operations turned up some exciting information in 1987. The USSR had just sent its fifteenth Meteor-2 weather satellite to a 600-mile-high orbit from the Northern Cosmodrome near Plesetsk on January 5, 1987, despite subzero temperatures which disrupted Europe and Asia that day.

Radio hobbyists in England reported radio signals from the new Meteor were interfering with transmissions from an older Meteor weather satellite in orbit and a secret Soviet military satellite known as Cosmos 1766 which used radar for ocean surveillance. All three satellites were transmitting on the same frequency.

On January 14, 1987, the USSR changed the new Meteor satellite frequency, but the British radio monitors still heard the old Meteor stuck on the old frequency, continuing to interfere with the military satellite.

ESA's Meteosat. The European Space Agency (ESA) has a series of weather satellites called Meteosat. They are controlled from ESA's Operations Centre at Darmstadt, Germany. Weather data is fed to national weather bureaus in Belgium, Denmark, Finland, France, Germany, Great Britain, Greece, Ireland, Italy, the Netherlands, Norway, Portugal, Spain, Sweden, Switzerland and Turkey.

Meteosat is a combined weather observer and radio relay in the sky. Its main payload is a sharp-eyed radiometer which takes infrared, water vapor and visible light pictures of the Earth. In infrared, it can resolve objects down to three miles. Visible light photos show objects down to 1.5 miles.

Meteosat takes pictures every half hour and transmits them to Darmstadt for processing. The processed images then are sent by radio back up to the satellite for relay down to the 16 countries using the weather data.

The Meteosat series started with a satellite orbited in 1977 and a second in 1981. The 1981 satellite was expected to work three years, but still is working in stationary orbit 22,245 miles above England. The 705 lb. cylinder is seven feet in diameter and ten feet tall.

New weather satellites have been launched to replace the 1977 and 1981 satellites. One Meteosat was launched in 1988 and an upgraded Meteosat known as MOP was launched in 1989.

Another MOP launch was scheduled for 1993. The new satellites have the added ability to relay weather charts and written weather reports to the national weather bureaus in the 16 countries.

Popular Weather Satellite Frequencies

Among radio hobbyists, weather satellites are the equivalent of shortwave utility monitoring or VHF-UHF scanner monitoring. Weather facsimile, popularly known as WEFAX, is the method of transmitting photographs and weather maps via radio and telephone. In weather satellites, it is referred to as APT. Many weather satellites in low orbits transmit in the 136-138 MHz band. Easily monitored low-orbit weather satellites include Chinese Feng Yun, Russian Meteor and Okean, and U.S. NOAA satellites. Frequencies in MHz:

Frequency	Satellite	Nation
137.300	Meteor 2-17	Russia
137.300	Meteor 3-03	Russia
137.300	Meteor 3-04	Russia
137.400	Meteor 2-13	Russia
137.400	Meteor 2-16	Russia
137.400	Okean 2	Russia
137.500	NOAA 10	USA
137.500	NOAA 12	USA
137.620	NOAA 9	USA
137.620	NOAA 11	USA
137.795	Feng Yun 1B when operating	China
137.850	Meteor 2-14	Russia
137.850	Meteor 2-15	Russia
137.850	Meteor 2-19	Russia
137.850	Meteor 3-01	Russia
1691.000	GOES	USA
1694.000	Meteosat	Europe

Japan's Sunflower. Japan's weather satellites are called Himawari or Sunflower. Four have been launched to orbit. The 715-lb. weather satellite Sunflower No. 4 was blasted off the Tanegashima Island launch pad September 6, 1989, on a three-stage H-1 rocket.

Sunflower No. 4 was sent to a stationary orbit 22,370 miles above New Guinea. The 1989 flight was the fourth successful launch in the Himawari series started in 1986 by Japan's National Space Development Agency (NASDA).

The $36.4 million satellite Sunflower No. 4 replaced the aging Sunflower No. 3 already in orbit sending weather pictures to ground stations. Sunflower No. 3 continued snapping cloud photos and transmitting them to Earth for another year.

NASDA operates the Tanegashima Island Space Center 616 miles southwest of Tokyo at the southern tip of Japan off Kagoshima Island on the southern tip of Kyushu, Japan's southernmost main island.

China's Wind and Cloud. The People's Republic of China launched its first weather satellite in 1988. Its first two weather satellites have been called Feng Yun No. 1 and Feng Yun No. 2, or Wind and Cloud No. 1 and No. 2. They were sent to polar orbits on Long March 4 rockets from Taiyuan space center in China's northern Shanxi province.

India's Insat. India has built and launched its own space satellites, and has had others build and launch satellites. The series called Insat, manufactured in the U.S., have been dual-purpose satellites to collect weather data as well as relay television and long-distance telephone service across India. Insat stands for India Satellite.

India's remote-sensing and Insat weather satellites have been launched by the U.S.,

Europe and the USSR. For instance, the Indian satellite Insat-1C was to have been ferried to space in a U.S. shuttle, but was switched to Ariane after the 1986 Challenger disaster. A European Space Agency Ariane 3 rocket ferried Insat-1C to space July 22, 1988.

The 2,618-lb. satellite was expected to have a working life of 10 years in a stationary orbit 22,500 miles above the equator. Unfortunately, part of the electrical power system in Insat 1-C broke down in space July 29, a week after riding Ariane 3 to orbit, according to the Indian Space Research Organization (ISRO). Then a part in one of two solar-powered electricity-generating systems malfunctioned. Only half of the satellite was connected to a power system that was operating properly.

Insat-2A and Insat-2B were scheduled for launch on Ariane in 1992. The India Remote Sensing (IRS) satellite was launched by the USSR March 17, 1988.

Other Earth-Observing Satellites

Remote-sensing earth-resources satellites, functioning very much like weather satellites, were invented in the U.S. in the 1960s to observe land masses and oceans, search for resources, record geological formations, monitor changes in environments and draw more accurate maps of Earth's surface.

Built to sense visible light, infrared light, microwave energy and X-rays from hundreds of miles above Earth, photos sent down by the satellites distinguish between soil and water, towns and pastures, ice and water in an ocean, wheat and corn in fields, and even distressed crops vs. healthy crops.

Best known has been the Landsat series of five American remote-sensing satellites launched between 1972 and 1984. Landsats transmit detailed mapping images of Earth's surface from polar orbits 400-600 miles above Earth. Landsats have resolved objects on the surface as small as 90 feet.

Landsats have been used to assess forests and rangelands, explore for oil and minerals, check vegetation and soil conditions, measure worldwide grain production, trace ocean currents, conduct geologic mapping, make environmental impact assessments, plan urban and rural growth, inventory major crops, monitor volcanoes, size-up reservoirs and lakes for dam-safety studies and spot ponds and rivers for water-resource planning.

Landsats 1 to 5, launched between 1972 and 1984, were built by the federal government and operated by the government until 1985 when U.S. President Ronald Reagan handed operation over to a private firm known as Earth Observation Satellite Co. (Eosat), a joint venture of Hughes Aircraft Co. and General Electric Co.

In 1992, President George Bush was asking Congress for funds to build a new Landsat, to be launched in the mid-1990s. To be known as Landsat 7, it would cost $250 million. The money would come from NASA and Dept. of Defense budgets.

Landsat	Launch	Landsat	Launch
Landsat 1	1972 Jul 23	Landsat 4	1982 Jul 16
Landsat 2	1975 Jan 22	Landsat 5	1984 Mar 1
Landsat 3	1978 Mar 5	Landsat 7	planned for 1995

The remote ocean sensing satellite SEASAT, launched in 1979, monitored surface wind speeds and temperatures, water-vapor content in the air, rate of rainfall, tides, storm surges, ice and currents. Seasat's radar measured wave heights from 3 to 65 feet with an altitude accuracy of 4 inches as it worked 106 days at an altitude of 500 miles, transmitting data to ground stations every 36 hours.

Other Earth-observation satellites in orbit above the planet include France's Spot photo satellites, Japan's Maritime Observation Satellites (MOS), European Space Agency's ERS satellite and numerous others launched by the former USSR, Europe and the U.S. India used Soviet rockets to send remote-sensing satellites to space from the USSR in 1987 and 1988. Canada wants to launch a radar remote-sensing satellite in 1994. China has a remote-sensing satellite and Brazil has one on the drawing board.

Spot. Similar Earth-observing satellites have been launched over the years by the USSR, the European Space Agency and Japan. A French Earth-observation satellite known as Spot-1 was shot to space in 1986 from the European Space Agency's French Guiana launch site to a circular orbit 500 miles above Earth. Spot-2 was lofted to orbit in 1990. Spot is owned by an organization combining private and government interests in Belgium, France and Sweden. The telescope attached to a camera in the satellite can resolve objects as small as 30 feet in size.

JERS-1. Japan launched its Earth Resources Satellite (JERS-1) in 1992. Ground controllers had problems unfurling the satellite's 40-ft. radar antenna in space.

Observation Satellite Frequencies

136.11 MHz	MOS-1B	Japan
2206.00 MHz	Spot	France
2265.50 MHz	Landsat	USA
2287.50 MHz	Landsat	USA
8150.00 MHz	JERS-1	Japan
8350.00 MHz	JERS-1	Japan
8400.00 MHz	Spot	France

Spot satellite photos show more detail than photos from Landsat. Spot has 30-ft. resolution versus Landsat's 90-ft. A camera in space with 30-ft. resolution will show an area as small as half a tennis court. JERS-1 shows objects as small as 60 feet. Russia beats all, however, selling photos with 18-ft. resolution.

News pix. Journalists around the world like to publish news photos made by Earth-resources satellites. Some satellite pictures have become famous. For instance, pictures made by Landsat 5 revealed to the world the Chernobyl nuclear power plant meltdown in 1986 in the former USSR. TV networks, newspapers and magazines bought Landsat and Spot pictures of Chernobyl and a Soviet naval base. Two years later, the U.S. Forest Service looked to Landsat photos for help battling forest fires in Yellowstone National Park. Landsat's infrared camera photographed the Valdez oil spill in Alaska in 1989. Spot satellite pictures showed airfields and a chemical weapons plant in Libya.

Weather Satellites Do Search And Rescue

SARSAT is shorthand for Search and Rescue Satellite-Aided Tracking System. COSPAS is a Russian acronym for Space System for Search of Vehicles in Distress. SARSAT/COSPAS is part of an international effort to find ships and planes in distress.

Ham catalyst. OSCAR 7, an amateur radio satellite built by ham operators in the 1970's, led to today's governmental SARSAT/COSPAS network. Under the direction of the Radio Amateur Satellite Corporation (AMSAT), OSCAR 7 was used in orbit in December 1975 to prove SARSAT/COSPAS would work. OSCAR 7 repeated low-power radio signals, received from the ground, down to Goddard Space Flight Center in Maryland, proving a weak uplink could provide accurate tracking to within two to four miles of an emergency site.

The governmental SARSAT/COSPAS system was set up in 1980 by Canada, France, the U.S. and USSR. It became operational in 1982. All U.S. fishing vessels and many other ships and planes around the world are required to carry SARSAT transmitters. The inexpensive, battery-powered devices are activated automatically in an emergency to fire distress signals up to space where they are intercepted by one or more of the search and rescue satellites and relayed to ground stations dotted around the globe.

SARSAT/COSPAS satellites are radio repeaters in space, housed in orbiting satellites. Since 1982, authorities credit the SARSAT/COSPAS satellites, and the rescue teams alerted on the ground, with having saved 1,200 persons around the world—mostly Americans—from shipwrecks, plane crashes and even a dog sled race which went awry. More than half a dozen government SARSAT/COSPAS satellites are listening in space today. A dozen countries support the network, including Canada, France, Great Britain, Norway, Russia, Sweden and the U.S.

ELTs and EPIRBs. While carrying out their assigned primary tasks in orbit, the SARSAT/COSPAS satellites monitor international radio distress frequencies for signals from emergency locator transmitters (ELTs). Airplane distress signals are sent by ELTs activated automatically upon impact. By analyzing the Doppler shift of an emergency radio signal heard and repeated by a SARSAT, ground crews can "fix" or pinpoint the ELT location to within two to four miles.

Emergency Position Indicating Radio Beacons, or EPIRBs for short, are small transmitters used to signal rescue services. Boaters in need of help, but out of two-way radio range, use EPIRB to summon aid. EPIRBs are legal only for Mayday emergencies in which a vessel is in danger of sinking or there is a medical emergency.

Class A and B EPIRBs transmit on the civilian VHF emergency frequency of 121.5 MHz and the military emergency frequency of 243.0 MHz. Class A automatically floats free and turns on. Class B is activated manually. Class A and B transmit weak line-of-sight signals, so they usually are found by airplanes which can see sufficient horizon to cover a sizable area of land or sea. Most aircraft are required to monitor the emergency frequency.

EPIRBs suffer from a 95 percent false-alarm rate. Only three percent of EPIRB signals come from true emergencies. EPIRBs do not identify themselves or the source of an emergency call.

SARSATs began listening to EPIRB frequencies in 1982. The satellites receive a signal from the ground and retransmit it to one of seven ground receiving stations, mostly in the Northern Hemisphere. Boaters in trouble in the Southern Hemisphere usually are out of luck. Their signals may be heard and repeated by satellite, but no ground station is beneath the satellite to receive the alert.

The latest EPIRB is referred to as "406" because it transmits on 406 MHz. Its signal is coded with an identification number, letting a ground station know who is in trouble and where to call to verify the emergency. False alarms are cut to five percent. Each 406 also transmits a weak signal on 121.5 MHz so planes and ships can home in as they get close.

The 406 transmitter is more powerful than older models. The SARSAT itself can calculate the emergency location to within one mile, making it easier to find the lost soul and cutting rescue costs. Unfortunately, 406 EPIRBs are very expensive compared with older models. Old styles sell around $200 while new types cost around $2,000.

SARSATs. Several U.S. and Russian satellites carrying SARSAT receivers have been launched to listen for distress signals from ships lost at sea and downed aircraft.

For instance, Russia's Cosmos 1861, launched in 1987 and stationed 600 miles above Earth, is a navigation satellite to help Russian fishing fleets and sea-going vessels locate themselves on the world's oceans. Besides its primary navigation payload and COSPAS gear, Cosmos 1861 also houses amateur radio equipment known as RS-10/11.

Also in 1987, the U.S. launched GOES 7, the first weather satellite in stationary orbit with a SARSAT receiver. It listens from 22,300 miles overhead. GOES-7's main job is transmitting cloud pictures and atmosphere temperatures to weather forecasters.

GOES-7 brought a new frequency of 406 MHz to SARSAT/COSPAS. Earlier SARSATs had been in polar orbits closer to Earth, with receivers tuned to the worldwide aircraft emergency beacon frequency of 121.5 MHz. The higher altitude of the stationary-orbit satellite, and the new 406 MHz frequency, often makes GOES-7 the first to hear a distress call and first to signal the alert. After an initial warning from GOES, polar satellites in lower orbits pinpoint the location of an emergency beacon.

NOAA-11, launched in 1988, listens from 500 miles overhead. It circles Earth every 102 minutes. As the planet rotates beneath, the satellite scans the entire globe every 12 hours. If it hears an emergency beacon, NOAA-11 relays the distress signal to a rescue service on the ground. NOAA 9, launched in 1984, and NOAA 10, launched in 1986, also had SARSAT receivers. The National Weather Service operates NOAA satellites from headquarters in Camp Springs, Maryland.

In 1989, the USSR launched another COSPAS aboard a civilian navigation satellite named Nadezhda, or Hope. Hope was orbiting at an altitude around 625 miles.

Sailors saved. Two Maryland men sailing a 35-ft. boat from Massachusetts to Bermuda owe their lives to a SARSAT which relayed their distress signals in a storm.

The boat was tacking 140 miles southeast of Nantucket December 17, 1987, when heavy wind and waves broke the mast. When the small craft took on more water than bilge pumps could handle, Chris Burtis of Annapolis and Troy Wilson of Bowie switched on their distress transmitter. A SARSAT overhead picked up the signal and relayed it to Scott Air Force Base in Illinois. The Air Force used the distress signal to determine the location of the sailboat. That information was sent to a Coast Guard air station on Cape Cod. The Coast Guard rescued Burtis and Wilson.

Flyer found. A brand-new NOAA 10 weather satellite, unlimbering during its first day of operation in space, saved the lives of Rory Johnston and three other Canadians when it found their small airplane crashed in a remote area of Ontario. Johnston's Cessna lost power during takeoff in poor weather, forcing an emergency landing on a lake.

The plane sank, nose down, in eight feet of water. With face cuts, bruised shoulder and dislocated wrist, Johnston swam 200 yards to shore where he found a canoe. He paddled back out to his sunken craft and brought his injured passengers ashore.

Meanwhile, the U.S. weather satellite NOAA 10 had been launched from California that very day, September 17, 1986. The SARSAT receiver had just been activated during the satellite's 76th orbit around the Earth. During its 90th revolution, while flying over Canada, the search and rescue radio heard Johnston's distress signal. NOAA 10 relayed the distress signal to Canadian rescue forces at Trenton.

At the same time, a Russian SARSAT also heard Johnston's beacon coming up from Ontario. NOAA 10 picked up the signal again on its 91st orbit. A pilot of a private plane also heard and reported Johnston's emergency beacon to authorities. The combination of reports from Russian and American SARSATs and the private pilot alerted rescue teams in Edmonton, Alberta. A four-engine C-130 Hercules plane with paramedics aboard took to the air in search of Johnston. Poor weather prevented the C-130 crew from spotting Johnston's plane that night. The next morning, when fog lifted, they found the Cessna and two medical technicians parachuted in to provide first aid. The SARSAT report had been so accurate the C-130 crew reported finding the distress beacon exactly where the SARSAT had said it was. After receiving first aid, Johnston and his three passengers were flown to an airstrip at Sachigo Lake, transferred to another plane and taken to Winnipeg, Manitoba, where they were hospitalized.

Life raft found. The 459-ft. freighter Independencia out of Port Washington, New

York, was 143 miles northeast of Kanton Island, halfway between Hawaii and Australia, enroute from Ecuador to China, when it sank in the Pacific Ocean December 11, 1991.

As its crew of 19, including 17 Pakistanis, one Kenyan and one Turk, abandoned ship between 3:20 and 5:15 a.m., a search and rescue team radioed the captain, asking him to turn on an Emergency Position Indicating Radio Beacon (EPIRB) in his life raft. The beacon transmitted a unique identification signal to outer space where it was heard by the umbrella of six polar-orbiting SARSAT/COSPAS satellites and tracked by Russian and American ground stations, leading rescuers to the Independencia. The entire Independencia crew of 19 was picked up from their 20-man life raft that afternoon by U.S. Coast Guard motor vessel Columbus Virginia and arrived at Honolulu harbor December 15.

How To Receive Weather Satellite Signals

Weather facsimile, or WEFAX, is the method for transmitting photos and weather maps via radio (wireless) and telephone (wired). In weather-satellite lingo, WEFAX is known as APT.

NOAA's National Environmental Satellite, Data and Information Service (NESDIS) receives WEFAX photos faxed from weather satellites. NESDIS draws maps over the pictures and transmits the resulting weather maps by radio, telephone and satellite.

Before launch to orbit, each weather satellite is assigned a transmitting frequency in a narrow segment of the radio spectrum known as very high frequency (VHF) or ultra high frequency (UHF). For instance, low-flying weather satellites such as NOAA and Meteor transmit on frequencies near 137.50 MHz. *(A list of satellite frequencies is on page 22.)*

To receive data and facsimile weather maps from low-flying, polar-orbiting satellites, you'll need an appropriate antenna, possibly a pre-amplifier to boost weak signals, a VHF or UHF receiver capable of tuning to the satellite's transmitting frequency, an interface device between receive and display, and a small computer with monitor or printer, or a special fax machine to display or print weather maps.

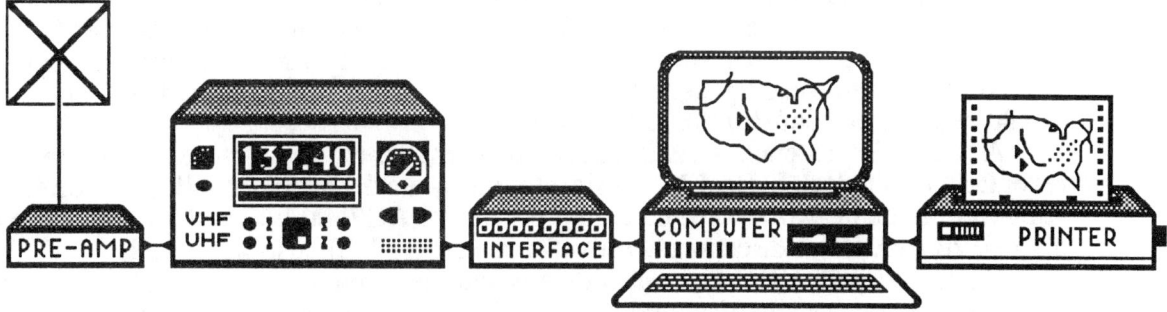

Pictures are easy to receive from low-flying, polar-orbiting satellites—such as NOAA, Meteor or Feng Yun transmitting APT at 136-138 MHz—because their radio signals are being transmitted from only 500-750 miles overhead.

They often can be received with modest equipment and a simple antenna. Sometimes, even a scanning monitor radio of the type used to receive local fire calls can receive signals from low-orbit weather satellites.

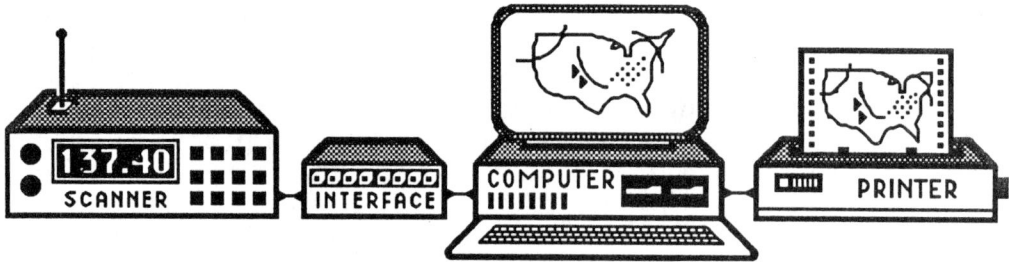

By comparison, stationary-orbit GOES weather satellites are much higher at 22,300 miles above Earth. Their signals become progressively weaker as they travel the longer distance, or path, to your receiving antenna. Intercepting signals from GOES and other high-flying, stationary-orbit weather satellites requires a large outside antenna and a more sensitive receiver capable of tuning to frequencies around 1690 MHz.

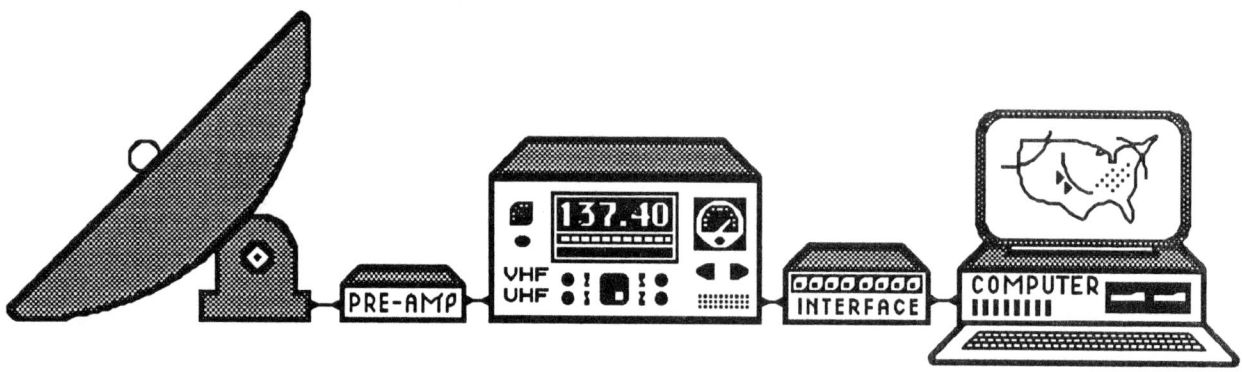

Antennas. As a low-orbit weather satellite revolves around the Earth, it passes over VHF receiving stations on the ground daily. Since such satellites are not far overhead, their strong signals require only simple antennas. An outdoor omni-directional antenna at a ground station will receive signals from a satellite when it is overhead, but the antenna should be tuned or "cut" for the 137 MHz band.

A "turnstile" antenna—designed after, but smaller than, those seen on housetops for reception of 88-108 MHz FM music stations—works well. Small omni-directional "discone" antennas work well, including the model AH7000 by ICOM and Radio Shack discone 20-013. Omni-directional "active" antennas for VHF also work well.

Do-it-yourself VHF antenna designs can be found in handbooks from many publishers. Antennas manufactured commercially for the amateur radio two-meter band at 144-148 MHz will work for 137 MHz satellite reception. A two-meter beam antenna can be rotated side to side and up and down to point at, or "track" a satellite as it passes overhead.

Monitoring stationary satellites, such as GOES and Meteosat at frequencies around 1690 MHz, is more difficult. Signals are very weak, so dish-shaped, 4-ft.-diameter receiving antennas and signal-boosting pre-amplifiers are common.

Frequency converters. Many weather-satellite hobbyists use frequency converters to move GOES signals received at 1690 MHz down to 137 MHz where it can be displayed with the same equipment used to receive low-flying polar-orbiting satellites.

Pre-amplifiers. While signals from polar-orbiting satellites often are strong on the ground, signals reaching the ground from stationary weather satellites are very weak. In a stationary satellite—such as GOES at an altitude of 22,300 miles above Earth—the transmitter puts out only a few watts of power and the signal has to travel thousands of miles from space to a ground receiving station.

To see weather satellite pictures clearly, especially from stationary-orbit satellites, you are likely to need a receiving pre-amplifier, or "pre-amp" for short, mounted at the antenna to boost incoming signal strength. Pre-amps using "GaAs-FET" technology are readily available at $50 or so, or make your own for less from widely-published do-it-yourself plans for amateur radio two-meter pre-amps.

You can use any low-noise receiver pre-amp designed for 137 MHz. The broadband 20 dB pre-amp model PR2 by Ramsey Electronics is in wide use. Other pre-amp manufacturers include Spectrum International and Quorum Communications. *(See the equipment manufacturer source list on pages 31-32.)* Do-it-yourself pre-amp designs can be found in handbooks published by the American Radio Relay League (ARRL).

Receivers. Weather satellite transmitters use frequency modulation (fm). However, their frequency deviation is wider than common narrow-band fm (nbfm) police and ham VHF radios. The deviation is narrower than the deviation of wide-band fm (wbfm) broadcast-band radios. Weather satellite fm transmitter deviation is nearly 20 KHz—four times the deviation of a typical police nbfm two-way radio and far less than the deviation of a wbfm music-broadcaster. The weather satellite receiver, therefore, should have an intermediate frequency (i.f.) passband between 30 and 60 KHz—37 KHz is best.

An nbfm VHF scanner or police monitor radio can be used to receive weather satellite pictures, but its intermediate frequency (i.f.) bandwidth of 15 KHz must be widened. A wbfm music radio could be used, but its intermediate frequency (i.f.) bandwidth of 150 KHz would have to be narrowed. An unmodified nbfm scanner would return distorted or "noisy" pictures, while the satellite signal would be too weak to be received on an unmodified wbfm music radio and pictures would be dark.

What to do? You could buy a receiver designed for weather satellite reception with an i.f. bandwidth filter of 50 KHz, or modify the i.f. of your old scanner to about 50-80 KHz. Commercial receiver model LABS 2000-B by Vanguard Labs is in wide use. Another widely used satellite receiver is made by Quorum Communications. *(See the equipment manufacturer source list on pages 31-32.)*

Warm Up The Soldering Iron

You might buy a special weather satellite receiver with an i.f. bandwidth filter of 50 KHz, or you could modify the i.f. of your old scanner to about 50-80 KHz.

One simple receiver modification, suggested in the March 1991 issue of the amateur radio magazine *73*, is removing the narrow i.f. filter in a police scanner and replacing it with a 0.01 µfd capacitor.

Computers. How do you know when to listen for signals coming down from low-orbit weather satellites? You might try listening from 1100z to 1400z and from 2100z to 0100z, but it's easier to use a computer or calculator to find the exact time. Computer programs require data called Keplerian elements, or "Keps" to calculate where a satellite is above Earth at any time.

Satellite tracking programs for popular personal computers are available from AMSAT (Radio Amateur Satellite Corporation), a worldwide organization of amateur radio operators building and using space satellites. Software also is available from Roy Welch, Rodger Mansfield, Paul Traufler, Bill Bard and others. *(Software source list, pages 31-32)*

Keplerian elements are available for many satellites in AMSAT publications and newsletters from other groups, but they can be obtained more quickly from computer bulletin boards (bbs). The orbits of satellites change constantly making Keplerian elements more than four weeks old inaccurate.

Many bbs's offer the latest Keps, including the giant commercial database service CompuServe at (800) 848-8990; the Celestial bbs at Fairborn, Ohio, (513) 427-0674; the Dallas Remote Imaging Group bbs at Carrollton, Texas, (214) 394-7438; the Canadian Space Society bbs at (416) 458-5907 and others.

Displaying pictures. Once you know when to listen, tune your receiver to the frequency of a satellite you wish to hear. *(Popular satellite frequencies, page 22)* Once tuned to a satellite at the right time, adjust your receiver to hear something like the sound of a ticking clock with a steady background of 2400 Hz audio tone carrying satellite pictures.

Audio from an APT satellite must be converted to a signal compatible with facsimile (fax) equipment. Among others, Vanguard and Quorum converters are used widely. Video data must be converted from analog to digital by a computer and software so satellite photos can be displayed on a computer's monitor screen. *(Source list, pages 31-32)*

Taggart. Ralph Taggart of Mason, Michigan, is an authority on reception of weather satellite images. His book, Weather Satellite Handbook, is available from the ARRL. Taggart's bbs is at (517) 676-0368. *(Supplier addresses, pages 31-32.)*

DRIG. In operation since 1984, a bbs operated by the Dallas Remote Imaging Group (DRIG) has become a major source of information on satellite tracking, frequencies, satellite pictures, remote sensing imagery, digital image and signal processing, and American, Chinese and Russian space programs. DRIG has 4000 members worldwide.

Using a personal computer and modem to connect with the DRIG bbs telephone line brings a caller into contact with more than two gigabytes of data storage holding thousands of images from NOAA weather satellites, amateur radio satellites, NASA's Voyager deep-space probes and many other Chinese, Russian and U.S. civilian and military satellites.

DRIG members, many with automated tracking and image-capture stations, upload pictures to the bbs. Image processing and display programs are available.

The bbs offers on-line tutorials, thousands of amateur radio and satellite telemetry programs, NORAD satellite Keplerian elements, weekly AMSAT bulletins and an ability to send E-mail to thousands of satellite tracking stations around the globe. The bbs telephone number for the general public and first-time users is (214) 394-7438.

First-time users who register while connected to the bbs receive information in the mail. Otherwise, prospective members may send a 9" x 12" envelope with six first class stamps to Dallas Remote Imaging Group *(address, page 31)* for a free membership package on tracking, frequencies, digital image processing, and a three-page primer on satellite photos. The DRIG office phone is (214) 394-7325. The fax line is (214) 492-7747.

An authoritative print-media source of weather satellite reception information is the quarterly Journal of the Environmental Satellite Amateur Users' Group (JESAUG) edited by Dallas Remote Imaging Group's Jeff Wallach.

WEFAX On TVRO

WEFAX maps sometimes can be found by tuning in those television broadcast satellites received in millions of homes with TVRO (television receive-only) gear. For instance, polar-orbiting NOAA and Russian Meteor satellites and GOES stationary satellites have been reported on Spacenet 3 transponder 19 in FDM (frequency division multiplexing) mode. A shortwave receiver must be used to unlock the weather photos.

To see and print satellite pictures you need a common TVRO satellite receiver system, a stable shortwave receiver, and a WEFAX decoder or demodulator feeding a personal computer system. Connect a 50-Ω coaxial cable from the baseband-output jack on the back of the satellite receiver to the antenna input connector on the back of the shortwave receiver. Feed audio output from the shortwave receiver into the WEFAX decoder or demodulator to display weather satellite pictures on the computer screen or copy them on a printer.

Set the satellite receiver to Spacenet 3 transponder 19. Tune the shortwave receiver across the 1-13 MHz range, searching for FDM signals. Meteor may be found at 1.568 MHz. When they work, GOES-West should be at 1.883 and GOES-East at 1.928 MHz. NOAA weather satellites also may be found by tuning the shortwave receiver.

Source List
Receivers, Pre-Amps, Antennas, TNC's, Interfaces, Computer Hardware, Software, WEFAX And Other Satellite And Shortwave Gear

A&A Engineering, 2521 W. La Palma, Unit K, Anaheim, California 92801.
Abrams, Clay, Software, 1758 Comstock Lane, San Jose, California 95124.
Accu-Weather, 619 West College Avenue, State College, Pennsylvania 16801.
AEA, P.O. Box 2160, 2006-196th Street SW, Lynnwood, Washington 98036.
Alden Electronics, 47 A Washington Street, Westborough, Massachusetts 01581.
Allegheny Microwave, 1135 Constitution Drive, Tarentum, Pennsylvania 15084.
American Meteorological Society, 45 Beacon Street, Boston, Massachusetts 02108.
American Radio Relay League (ARRL), 225 Main Street, Newington, Connecticut 06111.
American Weather Enterprises, P.O. Box 1383, Media, Pennsylvania 19063.
AMSAT, P.O. Box 27, Washington, D.C. 20044.
Applied Environmetrics, P.O. Box 241, Roslyn, Washington 98941.
Applied Neural Engineering, 3301 Coors Road NW # R-125, Albuquerque, New Mexico 87120.
Assn of State Climatologists, Western Regional Climate Center, P.O. Box 60220, Reno, NV 89506.
Associaton of American Weather Observers, P.O. Box 455, Belvedere, Illinois 61008.
ASTer Press/Storms, P.O. Box 466, Fort Collins, Colorado 80522.
Atlantic Sales, 3730 Nautilus Avenue, Brooklyn, New York 11224.
Azimuth, 3612 Alta Vista, Santa Rosa, California 95409.
Bard, Bill, 1732 74th Circle NE, St. Petersburg, Florida 33702.
Boone, C.F., Publishers, P.O. Box 1977, Sun City, Arizona 85372.
Bunyard, Joe, 1601 Lexington Street, Waco, Texas 76711.
Circuit Cellar Inc., Micromint Inc., 4 Park Street, Vernon, Connecticut 06066.
Cloud Chart, P.O. Box 21298, Charleston, South Carolina 29413.
Comar Electronics, Samuel Whites Estate, Medina Road, Crowes, Isle of Wight P031 7LF England
Commercial Weather Services Association, 655 15th Street NW #310, Washington, D.C. 20005.
Conversion Research, P.O. Box 535, Descanso, California 91916.
Coppola, Vince, 6 Robbin Road, Terryville, Connecticut 06786.
Dallas Remote Imaging Group, Jeff Wallach, P.O. Box 117088, Carrollton, Texas, 75011-7088.
Dartcomm, N. Hearn, Ferndale, Postbridge, Yelverton, Devon PL20 6SY, Great Britain.
Datametrics, 2575 South Bayshore Drive #8A, Coconut Grove, Florida 33133.
Davis Instruments, 3465 Diablo Avenue, Hayward, California 94545.
Design Works, 5 January Hills Road, Amherst, Massachusetts 01002.
Exacta-Weather, P.O. Box 931, Oneonta, New York 13820.
Feedback Instruments, Park Road, Crowborough, E. Sussex TN6 2QR, Great Britain.
Feely, Marty, 5273 Halifax Drive, San Jose, California 95130.
GTI Electronics, George Isleib, 1541 Fritz Valley Road, Lehighton, Pennsylvania 18235.
Hamtronics, Inc.,65-D Moul Road, Hilton, New York 14468-9535.
Hinds Instruments Inc., 5250 NE Elam Young Parkway, Hillsboro, Oregon 97124.
Horodner, Richard, 9961 SW 156 Terrace, Miami, Florida 33157.
Hurricane Research Service, P.O. Box 181032, Austin, Texas 78718.
ICOM America Inc., 2380 116th Avenue NE, Bellevue Washington 98004.
ICS Electronics Ltd., Rudford Ind. Estate, Ford, Arundel, W. Sussex BN18 0BD England
Imaging Publications, P.O. Box 66, Hubbardston, Massachusetts 01452.
International Traveler Weather Guide, P.O. Box 660606, West Sacramento, California 95866.

Source List continues on page 32

Source List continued from page 31

JAN Crystals, 2400 Crystal Dr, Fort Myers, Florida 33906-6017.
Johnson, Loren, P.O. Box 219, Cleveland, Minnesota 56017.
Kantronics, 1202 E. 23rd Street, Lawrence Kansas, 66046.
L'Weather Computer, Softworks Ltd, 803 12th Avenue NW, New Brighton, Minnesota 55112.
Mansfield, Rodger, Astronomical Data Service, 3922 Leisure Ln, Colorado Springs, Colorado 80917.
Martelec Comm. Systems, The Acorns, Wyck Ln, E. Worldham, Alton Hants GU34 3AW England.
McCallie Manufacturing Corp., P.O. Box 17721, Huntsville, Alabama, 35810.
MetraByte, 440 Myles Standish Road, Taunton, Massachusetts 02780.
Metsat Products, 1257 Glen Meadows Ln, East Lansing MI, 48823/P.O. Box 142, Mason, MI 48854
MFJ Enterprises Inc., P.O. Box 494, Mississippi State, Mississippi, 39762.
Microcomm, H. Paul Shuch, 14908 Sandy Lane, San Jose, California 95124.
Micromint Inc., 4 Park Street, Vernon, Connecticut 06066.
Multifax, David & Elmer Schwittek, 1659 Waterford Road, Walworth, New York 14568.
National Oceanographic Data Center, Users Services Branch, NOAA, Washington, D.C. 20235.
National Weather Association, 4400 Stamp Road, Room 404, Temple Hills, Maryland 20748.
Northern Marketing Concepts, 26 Woodlands Rd, Rillington Malton, N. Yorkshire Y0178LD, England
Oceansoft Inc, P.O. Box 1224, Largo, Florida 34649.
OFS Software, Jerry Dahl, 6404 Lakerest Court, Raleigh, North Carolina 27612.
Overview Systems, Tim Heffield, P.O. Box 130014, Sunrise, Florida 33313.
Quorum Communications, 1020 S. Main Street, Grapevine, Texas 76051.
R&D Electronics, The St. John Workshops, Margate, Kent CT9 1TE England.
Ramsey Electronics, 793 Canning Parkway, Victor, New York 14564.
RLD Research, McCloud, California 96057.
Royal Meteorological Society, 104 Oxford Road, Reading, Berkshire RG1 7LJ, England.
RTM Circuit Boards, 205 Elm Street, Van Horn, Iowa 52346-0400.
Satellite Data Systems Inc., P.O. Box 219, Cleveland, Minnesota 56017.
Schnedler Systems, 25 Eastwood Road, P.O. Box 5964, Asheville, North Carolina 28813.
Sensor Instruments, 41 Terrill Park Drive, Concord, New Hampshire 03301.
Softworks Inc., Fred Bartlett, Allentown, Pennsylvania.
Spectrum International, Inc., P.O. Box 1084, Concord, Massachusetts 01742.
Taggart, Ralph, 602 S. Jefferson, Mason, Michigan 48854.
Technical Software, Fron, Upper Llandwrog, Caernashon LL54 7RF England.
Texas Weather Devices, P.O. Box 309, Cresson, Texas 76035.
Texas Weather Instruments, 4925 Greenville Avenue #1302, Dallas, Texas 75206.
Tiare Publications, P.O. Box 493, Lake Geneva, Wisconsin 53147.
Timestep Weather Systems, Wickhambrook Market, CB8 8QA England.
Traufler, Paul E., 111 Emerald Drive, Harvest, Alabama 35749.
Universal Radio, Fred Osterman, 6830 Americana Parkway, Reynoldsburg, Ohio, 43068.
Vanguard Labs, 196-23 Jamaica, Hollis, New York 11423.
Weather Fax Inc., 52 Domino Drive, Concord, Massachusetts 01742.
Weather Gazette, P.O. Box 931, West Oneonta, New York 13820.
Weather Library, 1326 Sioux Street, Orange Park, Florida 32065.
Weather Modification Association, P.O. Box 8116, Fresno, California 93747.
Weather Network, 3760 Morrow Lane, Suite F, Chico, California 95928.
Weather School, 5075 Lake Road, Brockport, New York 14420.
WeatherDisc Associates, 4584 NE 89th, Seattle, Washington 98115.
Weathersat Ink, 4821 Jessie Drive, Apex, North Carolina 27502.
Weathertrac, P.O. Box 122, Cedar Falls, Iowa 50613.
Weatherwise Computers, 503 Walker Bluilding, University Park, Pennsylvania 16802.
Weatherwise Magazine, 1319 18th Street NW, Washington, D.C. 20036.
Welch, Roy D., 908 Dutch Mill Drive, Manchester, Missouri 63011.
Westwinds, 3540 76th Street, Caledonia, Michigan 49361.
White, Robert E., Instruments, 34 Commercial Wharf, Boston, Massachusetts 02110.
Wilsec Weather Instruments, 321 W. Grove Street, Greenville, Michigan 48838.
Wind & Weather, P.O. Box 1012, Mendocino, California 95460.
World WeatherDisc Associates Inc, 4584 NE 89th, Seattle, Washington 98115.

Shortwave WEFAX

NOAA's National Environmental Satellite, Data and Information Service (NESDIS) overlays maps on cloud photos faxed to Earth from weather satellites. The results are transmitted by WEFAX on high frequency (hf) bands to fishing boats and ships at sea.

WEFAX signals transmitted between 2 and 20 MHz are easy to monitor and display. You'll need a stable shortwave receiver with an appropriate antenna, a terminal node controller (tnc) and a small computer or fax device to display or print weather maps.

There are many shortwave receivers suitable for WEFAX reception, including those manufactured by Kenwood, Icom, Yaesu, Japan Radio Company, Sony and Panasonic.

Terminal node controllers include models by Kantronics, AEA, MFJ and others. For instance, AEA's tnc model PK-232 displays black-and-white fax images, including weather maps and GOES satellite photos received on shortwave frequencies. You also can use the MFJ 1278 with MFJ Multicom software to display maps. Terminal node controllers work with all computer brand names through standard RS-232 connections or "ports." WEFAX display software is available for Apple and Macintosh, IBM PCs and compatibles, Commodore and Atari models and other systems. Do-it-yourself hf antenna designs can be found in handbooks from many publishers. Antennas manufactured for hams and shortwave listeners will capture WEFAX signals. *(Suppliers of receivers, preamps, antennas, tnc's, interfaces, computer hardware and software, and other shortwave WEFAX gear, pages 31-32.)*

Weather facsimile is transmitted on the following frequencies. Depending upon type of receiver used, listeners may have to compensate for single sideband transmission. For example, 8.080 MHz would be 8.078 upper sideband (USB).

Shortwave WEFAX Frequencies

MHz	From
2.122	Hawaii, 0600-1600z
2.6185	Bracknell, Great Britain, 1800-0600z, April 1-September 30
2.81385	Northwood, Great Britain, 1630-0730z, September 30-March 31
2.815	Moscow, 1800-0055z
3.235	Frobisher Bay, California, 1000-2200z, July 1-October 15
3.2895	Bracknell, Great Britain
3.357	NOAA WEFAX 24 hours, station NAM, Norfolk, Virginia
3.377	Ankara Turkey, 1600-0030z

3.3775	Guam
3.43685	Northwood, Great Britain
3.520	Belgrade Yugoslavia, 1700-0700z
3.650	Madrid, 0400-1700z
3.855	Hamburg, Germany, 0900-1000z
4.0375	Norrkoping, Sweden
4.0475	Paris
4.223	Delaware, WLO
4.24785	Northwood, Great Britain
4.268	Esquimalt, British Columbia, Canada
4.271	Halifax, Nova Scotia
4.296	Kodiak, Alaska
4.322	Monsanto, 0635-1700z
4.346	San Francisco, 0100-1500z
4.5167	Khabarovsk
4.610	Bracknell, Great Britain
4.704	Rota, Spain, station AOK
4.768	Rome
4.782	Bracknell, Great Britain
4.793	Washington, DC
4.855	Hawaii
4.975	Guam
5.093	Sofia, Bulgaria
5.206	Athens, Greece, 2000-0800z
5.355	Moscow
5.800	Belgrade, Yugoslavia
5.850	Copenhagen, Denmark, 0030-1005z
6.185	Madrid 0400-1700z
6.330	Halifax, Nova Scotia, station CFH
6.4365	Northwood, Great Britain
6.790	Ankara, Turkey, 0500-1400z
6.850	Mobile, Alabama, station WLO
6.872	Buenos Aires, station LRB79, news photos facsimile
6.901	Norrkoping, Sweden
6.944	Vancouver, British Columbia, station CKN
6.968	Esquimalt, British Columbia, Canada
7.417	Rota, Spain
7.475	Khabarovsk
7.530	Boston, Massachusetts, station NMF
7.5875	Dakar, 2000-0830z
7.645	Guam
7.710	Frobisher Bay,, California, 1000-2000z, July 1-October 15
7.750	Moscow
7.770	Hawaii
7.880	Hamburg, Germany, 0900-1000z
8.018	Helsinki, Finland, 0740z
8.040	Bracknell, Great Britain
8.075	Norrkoping, Sweden
8.080	NOAA WEFAX 24 hours, station NAM, Norfolk, Virginia
8.100	Athens, Greece
8.146	Rome
8.185	Paris
8.260	Monsanto, 0635-1700z
8.457	Kodiak, Alaska
8.494	Alaska

8.4945	Northwood, Great Britain
8.502	Boston, MA, 1600z, ice March-July
8.502	Norfolk VA
8.646	San Diego, CA
8.682	San Francisco
8.682	San Francisco,, California, station NMC
9.092	Brentwood NY, 0712-1212z
9.157	Mobile, Alabama, station WLO
9.1575	Mobile, Alabama, 0250-2030z
9.203	Bracknell, Great Britain
9.230	Khabarovsk
9.360	Copenhagen, 0005-1850z
9.3895	Brentwood NY, 0712-1212z
9.440	Hawaii
9.875	Rota, Spain
9.890	Halifax, Nova Scotia
9.9825	Honolulu
10.123	Cairo
10.185	Washington, DC
10.250	Madrid, Spain, 0400-1700z
10.255	Guam
10.535	Halifax, Nova Scotia, station CFH
10.677	Buenos Aires, station LRN2
10.865	NOAA WEFAX 24 hours, stn NAM, Norfolk, Va., schedule at 2400z
11.035	Brentwood NY
11.090	Hawaii
12.201	Washington, DC
12.710	Japan
12.730	San Francisco
12.750	Boston
13.510	Halifax, Nova Scotia
13.751	London, news photos facsimile
13.862	Hawaii
14.6715	Washington, DC
14.828	Pearl Harbor, Hawaii, station NPM
16.410	NOAA WEFAX 0900-2100z, station NAM, Norfolk, Virginia
17.1512	San Francisco
17.4105	San Diego
17.4475	Mobile, Alabama
17.585	Rota, Spain, station AOK
17.670	Buenos Aires, station LQZ67, news photos facsimile
18.431	Buenos Aires, station LRO83, news photos facsimile
20.015	NOAA WEFAX 1200-2100z, station NAM, Norfolk, Virginia
20.246	Cairo
22.542	Tokyo, Japan, station JJC, news photos facsimile

Shortwave WEFAX maps are black and white, while GOES weather satellite photos are continuous gray-scale. WEFAX is transmitted on shortwave at 120 lines per minute. A tnc designed for WEFAX will receive maps clearly and give a good view of cloud cover in GOES photos without gray-scale. Here is the WEFAX transmission schedule:

UTC	Data
0000/1200	Schedule of Transmissions / NMC Boundary Layer Analysis
0015/1215	850MB. Height/Temperature/Wind 36 Hour Prog
0030/1230	500MB. Height/Temperature/Wind 36 Hour Prog

0045/1245	500MB. Height/Temperature/Wind 48 Hour Prog
0100/1300	Surface Pressure 36 Hour Prog (South Atlantic)
0115/1315	Surface Pressure 48 Hour Prog (South Atlantic)
0130/1330	500MB. Pressure 36 Hour Prog (South Atlantic)
0145/1345	500MB. Pressure 48 Hour Prog (South Atlantic)
0200/1400	GOES satellite channel 02 (Full Disk)
0215/1415	NMC Extended Surface / NMC 200MB Ht Analysis
0230/1430	Open Period
0240/1440	RAFC SIG weather 12 Hour Prog (FL250-FL600)
0250/1450	NMC 36 Hour 500MB Height/Isotach Prog
0300/1500	NMC 24 Hour 500MB Height Change
0315/1515	NEOC 36 Hour Prog Blend
0330/1530	National Weather Service Radar Summary
0345/1545	Surface Pressure Analysis
0400/1600	500MB. Pressure Analysis (South Atlantic)
0415/1615	FNOC Preliminary Surface Analysis (North Atlantic)
0430/1630	Surface Trop Pressure/Wind Analysis
0445/1645	Open Period (Tropical Warnings)
0500/1700	NMC NGM 24 Hour Prog
0515/1715	GOES satellite channel 14 (Gulf of Mexico)
0530/1730	Bracknell 24 Hour Surface Prog
0545/1745	GOES satellite channel 15 (North Atlantic)
0600/1800	NMC NGM 48 Hour Prog
0615/1815	NMC Radar Summary / NEOC Sea Ht Analysis
0630/1830	850MB. Height/Temperature/Wind Analysis
0645/1845	700MB. Height/Temperature/Wind Analysis
0700/1900	500MB. Height/Temperature/Wind Analysis
0715/1915	300MB. Height/Temperature/Wind Analysis
0730/1930	Surface Pressure/Wind 12 Hour Prog
0745/1945	GOES satellite channel 14 (Gulf of Mexico)
0800/2000	Surface Pressure/Wind 24 Hour Prog
0815/2015	Surface Pressure/Wind 36 Hour Prog
0830/2030	NMC 24 Hour Sig Weather
0840/2040	NMC 36/48 Hour Sig Weather
0850/2050	Bracknell 48 Hour Sig Wave / NEOC 84 Hour Prog Blend
0905/2105	Sig Wave Height 24 Hour Prog
0920/2120	Surface Pressure Analysis Preliminary (06&18Z)
0935/2135	200MB Height/Temperature/Wind 24 Hour Prog
0950/2150	Open Period (Tropical Warnings)
1000/2200	NEOC 12 Hour High Wind/Sea Warnings
1015/2215	850MB. Height/Temperature/Wind 24 Hour Prog
1030/2230	700MB. Height/Temperature/Wind 24 Hour Prog
1045/2245	500MB. Height/Temperature/Wind 24 Hour Prog
1100/2300	400MB. Height/Temperature/Wind 24 Hour Prog
1115/2315	GOES satellite channel 15 (North Atlantic)
1130/2330	300MB. Ht/Temp/Wind 24 Hour Prog
1145/2345	Freezing Level 24 Hour Prog

Hurricane Hunters

Death-defying Hurricane Hunters fly into the heart of major storms to supply the U.S. National Hurricane Center at Coral Gables, Florida, with critical measurements meteorologists need to judge size, strength, pinpoint location and predict storm track.

NOAA's hurricane-hunting aircraft are WP-3D Orion planes outfitted with special gear to collect raw weather data. They are flown directly into upper levels of hurricanes churning across the Caribbean Sea and Atlantic Ocean.

The Hunters stay in touch with the National Hurricane Center at Coral Gables by voice and other modes. Most are direct transmissions on shortwave, but sometimes voice and data communications from the aircraft are routed through NASA's ATS-3 satellite.

Some aircraft transmit telemetry data to a working Geostationary Operational Environmental Satellites (GOES). When the satellites are working, GOES-East is positioned over the equator at 75 degrees west and GOES-West at 135 degrees west. Yet another system—Hurricane Hunter Aircraft Packet Project (HHAPP)—uses an airborne amateur radio packet-radio digipeater, transmitting 10 watts of power in the two-meter amateur radio band.

Hurricane Hunters communications can be heard on the hf, vhf and uhf bands.

MHz	Use
3.407	air-to-ground reconnaissance aircraft communications, usb
4.701	air-to-air backup reconnaissance aircraft communications, usb
5.562	air-to-ground reconnaissance aircraft communications, usb
6.673	air-to-ground reconnaissance aircraft communications, usb
8.876	air-to-ground reconnaissance aircraft communications, usb
10.015	air-to-ground reconnaissance aircraft communications, usb
11.398	air-to-ground reconnaissance aircraft communications, usb
13.267	air-to-ground reconnaissance aircraft communications, usb
21.937	air-to-ground reconnaissance aircraft communications, usb
123.050	air-to-air primary reconnaissance aircraft communications, am
136.370	air-to-National Hurricane Center via NASA's ATS-3 satellite
145.010	HHAPP packet radio digipeater, data communications, 10 watts, nbfm
304.800	air-to-air secondary reconnaissance aircraft communications
405.000	air-to-GOES satellite data uplink for NOAA P-3 aircraft
1691.000	GOES satellite-to-ground data communications

Identifying Hurricane Hunters

aircraft identification	NOAA + last digit of aircraft registration number or A + last 3 digits of tail number
mission/depression number	depression number or XX if not a depression or greater
storm name	actual storm name

examples:
AF985 01XX INVEST	Air Force aircraft 985 on first mission to investigate a suspect area.
AF987 0503 CYCLONE	Air Force aircraft 987 on the fifth mission to depression number 3.
NOAA2 0701 LOUIS	NOAA aircraft 42RF on the seventh mission to fix depression number 1, which has acquired the name Louis.
observation numbering and content:	The first weather observation will have as remarks an ICAO four-letter departure station identifier, time of departure, and estimated time of arrival at the coordinates or storm. AF966 0308 EMMY OB 01 97779 TEXT TEXT...DPDT KBIX AT 10/2100Z ETA 31.5N 75.0W AT 11/0015Z

Hurricane Disaster Potential

Hurricanes are categorized from 1 to 5 on the Saffir-Simpson scale. Pressure is in inches. Wind speeds are in miles per hour. Storm surge is in feet.

Category	Pressure	Winds	Surge	Damage
1	28.94	74-95	4-5	**Minimal.** Primarily trees, foliage, unanchored mobile homes. No real damage other structures. Some small craft torn from moorings.
2	28.50	96-110	6-8	**Moderate.** Some window, door, roofing damage, trees down. Small craft in unprotected anchorages torn from moorings. Some evacuation shoreline, low-lying islands.
3	27.91	111-130	9-12	**Extensive.** Large trees down. Some structural damage small buildings. Mobile homes destroyed. Serious coastal flooding. Wind, waves destroy many small structures near coast. Almost all small boats torn from moorings.
4	27.17	131-155	13-18	**Extreme.** Residences extensive roof damage. Terrain flooded 10 ft. above sea level. Requires massive coastal evacuation. Rising water cuts escape routes 3 to 5 hours before center arrives.
5	27.16	156-up	19-up	**Catastrophic.** Complete failure of roofs on residences and many commercial structures. Small buildings overturned or blown away. Massive evacuation from low ground within 5 to 10 miles of coast.

NOAA uses several formats to transmit data from Hurricane Hunters to Coral Gables. The codes include VORTEX for voice communications, RECCO for telemetry and AFOS for the HHAPP packet-radio system.

VORTEX Voice Message Codes

Vortex fix	Location of surface and/or flight level center of a tropical or subtropical cyclone obtained by reconnaissance aircraft penetration
A	hrs Z, date and time
B	deg N/S, latitude of vortex fix; deg E/W, longitude of vortex fix
C	mb/m, millibars/meters, minimum height at standard level
D	knots, estimate of maximum surface wind observed
E	deg/nm, bearing/range from center of maximum surface wind
F	deg/kt, maximum flight level wind near center
G	deg/nm, bearing and range from center of maximum surface wind
H	mb, minimum sea level pressure computer from dropsonde or extrapolated from within 1500 feet of sea surface
I	C/m, maximum flight level temperature/pressure alt outside eye
J	C/m, maximum flight level temperature/pressure alt inside eye
K	C/c, dew point temperature/sea surface temperature inside eye
L	eye character: closed wall, poorly defined, open SW, etc.
M	Eye shape/orientation/diameter: Code eye shapes as: C is circular; CO is concentric; E is elliptical. Transmit orientation of major axis in tens of degrees, i.e., 01 is 010 to 190; 17 is 170 to 350. Transmit diameter in nautical miles. Examples: C8 is circular eye 8 miles in diameter. E09/15/5 is elliptical eye, major axis 090-270, length of major axis 15 nm, length of minor axis 5 nm. CO8-14 is concentric eye, diameter inner eye 8 nm, outer eye 14 nm

N   deg N/S, confirmation of fix: coordinates & time; deg E/W; time Z
O   fix determined by/fix level: fix determined by: 1 penetration; 2 radar; 3 wind; 4 pressure; 5 temp. Fix level (indicate surface center if visible; indicate both surface & flight level centers only when same): 0 is surface; 1 is 1500 ft.; 8 is 850 mb; 7 is 700 mb; 5 is 500 mb; 4 is 400 mb; 3 is 300 mb; 2 is 200 mb; 9 is other.
P   nm, navigation fix accuracy /meteorological accuracy
Q   remarks

RECCO Reporting Codes

Group	Format	Group	Format
1	9XXX9	13	1K(n)N(s)N(s)N(s)
2	GGggi(d)	14	Ch(s)h(s)H(t)H(t)
3	YQL(a)L(a)L(a)	15	Ch(s)h(s)H(t)H(t)
4	L(o)L(o)L(o)Bf(c)	16	Ch(s)h(s)H(t)H(t)
5	h(a)h(a)h(a)d(t)d(a)	17	4ddff
6	ddfff	18	6W(s)S(s)W(d)d(w)
7	TTT(d)T(d)w	19	6W(s)S(s)W(d)d(w)
8	/jHHH	20	7I(r)I(t)S(b)S(e)
9	1k(n)N(s)N(s)N(s)	21	7h(i)h(i)H(i)H(i)
10	Ch(s)h(s)H(t)H(t)	22	8d(r)d(r)S(r)O(e)
11	Ch(s)h(s)H(t)H(t)	23	8E(w)E(l)c(e)i(e)
12	Ch(s)h(s)H(t)H(t)	24	9V(i)T(w)T(w)T(w)

Group 1 9XXX9 (9's are non-significant group indicators)
222   Sec One Observation without radar capability
555   Sec Three (Intermediate) observation with or without radar capability
777   Sec One Observation with radar capability

Group 2 GGggi(d)
GGgg Time of observation (hours/minutes UTC)
i(d) Dew point indicator:
    0 No dew point capability/aircraft below 10 km
    1 No dew point capability/aircraft at or above 10 km
    2 No dew point capability/aircraft below 10 km and flight level temp -50 C or colder
    3 No dew point capability/aircraft at or above 10 km and flight level temp -50 C or colder
    4 Dew point capability/aircraft below 10 km
    5 Dew point capability/aircraft at or above 10 km
    6 Dew point capability/aircraft below 10 km and flight level temp -50 C or colder
    7 Dew point capability/aircraft at or above 10 km and flight level temp -50 C or colder

Group 3 YQL(a)L(a)L(a)
Y   day of week, Sunday=1
Q   octant:

0	0-90 [deg] west, northern		5	0-90 west, southern
1	90-180 west, northern		6	90-180 west, southern
2	180-90 east, northern		7	180-90 east, southern
3	90-0 east, northern		8	90-0 east, southern
4	not used			

L(a)L(a)L(a) latitude, degrees and tenths

Group 4 L(o)L(o)L(o)Bf(c)
L(o)L(o)L(o) longitude, degrees and tenths; hundreds omitted from 100 - 180 degrees

B   Turbulence: f(c)   Flight conditions:
    0 None   0 In the clear
    1 Light turbulence   8 In and out of clouds
    2 Moderate turbulence in clear air, infrequent   9 In clouds all the time (continuous IMC)
    3 Moderate turbulence in clear air, frequent   / Impossible determine due to darkness or other cause
    4 Moderate turbulence in cloud, infrequent
    5 Moderate turbulence in cloud, frequent
    6 Severe turbulence in clear air, infrequent
    7 Severe turbulence in clear air, frequent
    8 Severe turbulence in cloud, infrequent

    9   Severe turbulence in cloud, frequent

Group 5 h(a)h(a)h(a)d(t)d(a)
h(a)h(a)h(a) pressure altitude of aircraft reported to the nearest decameter
d(t) Type of wind: d(a) Method of obtaining wind:
       0   Spot wind   0   Doppler radar or inertial systems
       1   Average wind   1   Navigation equipment and/or techniques
       /   No wind reported   /   Navigator unable to determine wind or wind not compatible

Group 6 ddfff
dd   Wind direction at flight level; tens of degrees true
fff   Wind speed at flight level, knots

Group 7 TTT(d)T(d)w
TT   Temperature (whole degrees C)
T(d)T(d) Dew point (whole degrees C) If negative, 50 added to absolute value with hundreds, if any, omitted
w Present weather:
   0   Clear   5   Drizzle
   1   Scattered (trace to 4/8 cloud coverage) 6   Rain, continuous/intermittent precip from stratiform clouds
   2   Broken (5/8 to 7/8 cloud coverage)   7   Snow or rain and snow mixed
   3   Overcast/undercast   8   Shower(s)continuous/intermittent precip from cumuliform clouds
   4   Fog, thick dust or haze   9   Thunderstorm(s)
                                                   /   Unknown for any cause including darkness

Group 8 /jHHH
/   Indicator (insignificant)
j   Index to HHH:
     0   Sea level pressure in whole millibars; thousands figure omitted
     1   Altitude 200 mb surface in geopotential decameters; thousands figure omitted
     2   Altitude 850 mb surface in geopotential meters; thousands figures omitted
     3   Altitude 700 mb surface in geopotential meters; thousands figures omitted
     4   Altitude 500 mb surface in geopotential decameters
     5   Altitude 400 mb surface in geopotential decameters
     6   Altitude 300 mb surface in geopotential decameters
     7   Altitude 250 mb surface in geopotential decameters; thousands figure omitted
     8   D, Value in geopotential decameters; if negative, 500 is added to HHH
     9   No absolute altitude available or geopotential data not within +/− 30 meters/4mb accuracy
HHH Geopotential height/D Value or SLP per index j

Group 9 1k(n)N(s)N(s)N(s)
1   Indicator (insignificant)
k(n)   Number of cloud layers
N(s)N(s)N(s) Amount of clouds

Group 10 Ch(s)h(s)H(t)H(t)
C   Cloud type:
      0   Cirrus   5   Nimbostratus
      1   Cirrocumulus   6   Stratocumulus
      2   Cirrostratus   7   Stratus
      3   Altocumulus   8   Cumulus
      4   Altostratus   9   Cumulonimbus
      /   Cloud type unknown due to darkness or other analogous phenomena

---

Legal Reminder

Two-way point-to-point communications are different from one-way broadcasts intended for dissemination throughout the general public. The privacy of point-to-point communication is protected by federal law. Listeners are forbidden to reveal or repeat to anyone the content of any two-way radio communication, except amateur radio communications. The prohibition includes reporting anything heard to news media.

VOLMET And Aviation Weather

VOLMET, a French acronym for Aviation Weather, is a chain of shortwave stations around the globe broadcasting a regular schedule of forecasts and weather data for pilots of international airplane flights. The VOLMET net uses usb on several hf frequencies.

The list shows the region covered, station locations and frequencies in MHz. The abbreviation hr stands for hour. For instance, hr+30 is 30 minutes past the hour. Broadcasts are about five minutes long.

Region **Station**

Pacific Pacific region frequencies: 2.863, 6.679, 8.828, 13.282 MHz

HONOLULU RADIO   weather time: hr+20, hr+50
Wake, Guam, Kahului, Wake, Kahului, Hilo

OAKLAND RADIO   weather time: hr+5, hr+35
San Francisco, Los Angeles, Seattle, Portland, Ontario, Los Angeles, San Francisco, Seattle

TOKYO RADIO   weather time: hr+10, hr+40
New York, Tokyo, Chitose, Nagoya, Osaka, Fukuoka, Seoul, New Tokyo

HONG KONG RADIO   weather time: hr+15, hr+45
Cheung Chau, Naha Okinawa, Taipei, Kao Hsiung, Manila, Macao, Hong Kong

AUCKLAND RADIO   weather time: hr+20, hr+50

ANCHORAGE RADIO   weather time: hr+25, hr+55
Fairbanks, Cold Bay, King Salmon, Shemya, Vancouver, Anchorage, Cold Bay

Europe European region frequencies: 2.998, 3.413, 5.640, 6.580, 8.957, 11.378, 13.264 MHz

BEN GURION RADIO   weather time: hr+5-15, H+35-45
Ben Gurion, Ramat David, Elat, Athens, Ankara, Istanbul, Tel Aviv

PRAGUE RADIO   weather time: hr+15-25, hr+45-55
Prague, Bratislava, Brno, Vienna, Berlin, Warsaw, Munich, Budapest, Moscow

PARIS RADIO   weather time: hr+25-35, hr+55-5
Paris-Le Bourget, Paris-Orley, Tours, Reims, Bordeaux, Toulouse, Marseille, Lyon, Nice, Algiers

North Atlantic day frequencies: 5.592, 8.870, 13.270   night frequencies: 2.905, 5.592, 8.870 MHz

SHANNON AERADIO   weather time: hr+0-25, hr+30-55
April-October operations: 2200-1000z   November-March operations: 1800-1200z
Shannon, Madrid, Prestwick, London Heathrow, Amsterdam, Oslo, Copenhagen, Athens, Paris

Southeast Asia Southeast Asian frequencies: 2.965, 3.458, 5.673, 6.676, 8.849, 11.387, 13.285 MHz

SYDNEY RADIO   weather time: hr+0, hr+30
Sydney, Brisbane, Darwin, Adelaide, Alice Springs

CALCUTTA RADIO   hr+5, hr+35
limited operation freqs: 6.676 (24 hrs); 3.458 (1200-0300z); 11.387 (0300-1300z)
Gaya, Delhi

BANGKOK RADIO   hr+10, hr+40
limited operation frequencies: 6.676, 11.387 (2310-1145z); 3.458, 6.676 (1210-2245z)
Rangoon, Saigon, Bangkok

KARACHI RADIO   hr+15, hr+45
limited operation frequencies: 3.458, 6.676 (1610-0140z); 6.676, 11.387 (0140-1510z)
Nawabshah, Jiwani

SINGAPORE RADIO   hr+20, hr+50

limited operation frequencies: 6.676 (1230-2230z) and 11.387 (2230-1230z)
Kuala Lumpur, Kota Kinabalu, Penang

	BOMBAY RADIO	hr+25, hr+55

limited operation frequencies: 3.458, 6.676, and 11.387
Ahmadabad, Calcutta

Middle East — Middle Eastern frequencies: 2.956, 5.589, 8.945, 11.393 MHz

	BAGHDAD RADIO	weather time: hr+0, hr+30

Baghdad, Basrah

	TEHRAN RADIO	weather time: hr+5, hr+35

Abadan, Tehran

	BAHRAIN RADIO	weather time: hr+10, hr+40

Dhahran, Kuwait

	BEIRUT RADIO	weather time: hr+15, hr+45

Nicosia, Cairo, Amman, Jeddah, Athens

	YESILKOY RADIO	weather time: hr+25, hr+55

Istanbul, Ankara, Athens

Atlantic Ocean — Atlantic region freqs: 2.905, 3.485, 5.592, 6.604, 8.870, 10.051, 13.270, 13.276MHz

	NEW YORK RADIO	weather time: hr+0, hr+30
	GANDER AERADIO	weather time: hr+20, hr+50

North Africa — North African frequencies: 2.860, 3.404, 5.499, 6.538, 8.852, 10.057, 13.261 MHz

	ALGER RADIO	weather time: hr+0, hr+10, hr+40

Alger, Annaba, Oran, Tripoli, Benghazi

West Africa — West African frequencies: 2.860, 3.404, 5.499, 6.538, 8.852, 10.057, 13.261 MHz

	DAKAR AERADIO	weather time: hr+15-25, hr+45-55

Dakar, Lome, Bamako, Lagos, Conakry, Abidjan, Niamey, Libreville

Indian Ocean — Indian Ocean region frequencies: 2.881, 5.601, 10.087, 13.279 MHz

	TANANARIVE RADIO	weather time: hr+25, hr+55

Tananarive, Saint Denis, Diego Suarez

FAA, Military And Coast Guard Weather

Worldwide, regional and local transmissions of forecasts and weather data to military forces in the air and at sea. Also, communications to and from Federal Aviation Administration aircraft and Coast Guard emergency flights.

The list shows some transmitter locations, coverage areas and transmit frequencies in MHz.

Service	Location	Coverage	MHz
military	Halifax	regional, global	3.046, 6.746, 11.249
military	Lahr	regional, global	5.690, 6.753, 13.231
military	Edmonton	regional, global	6.705, 6.753, 15.035
military	Trenton	regional, global	6.705, 6.753, 15.035
military	St. Johns	regional, global	6.705, 15.035
military	military flights	local, regional	272.700, 342.50
Coast Guard	Coast Guard air and sea rescue	local, regional	287.80, 237.90
	aircraft emergency	local, regional	121.50, 243.00
FAA	Federal Aviation Administration	local, regional	342.50, 344.60, 239.80
local airports	airport advisory	local, regional	123.60
U.S.	aviation weather	nationwide	122.00

ATIS And AWOS

Automated Terminal Information Service (ATIS) and Automated Weather Observing System (AWOS) are very weak radio signals emanating from most airports, providing continuous weather reports, conditions and other information about the immediate surroundings of the airfields.

The very-low-power transmitters, operating in the 108-136 MHz air band, are intended to have a short range of three miles at ground level.

Countless ATIS or AWOS stations are located at airports across the nation. Here are some frequency examples of transmitters reported in Delaware, Maryland and Virginia:

ATIS	Airfield	AWOS	Airfield
123.950	Delaware, New Castle County	118.375	Virginia, Portsmouth
108.400	Maryland, Philipps Army Air Field	118.425	Virginia, Charlottesville
		127.525	Virginia, Manassas

Weather data-gathering gear usually is located adjacent to runways. AWOS broadcasts an updated forecast every minute. It reports the name of the airport, universal time (zulu), and data in one of four patterns:

Pattern	Content
AWOS-A	altimeter setting
AWOS-1	altimeter setting, winds, temperature, dewpoint, density
AWOS-2	altimeter setting, winds, temperature, dewpoint, density altitude, visibility
AWOS-3	altimeter setting, winds, temperature, dewpoint, density altitude, visibility, cloud ceiling

Legal Reminder

Two-way point-to-point communications are different from one-way broadcast communications intended for dissemination throughout the general public. The privacy of point-to-point communication is protected by federal law. Listeners are forbidden to reveal or repeat to anyone the content of any two-way radio communication, except amateur radio communications. The prohibition includes reporting to any news media.

Maritime Weather Services

Weather forecasts and data from the National Weather Service can be found on long-range shortwave frequencies between 2 and 23 MHz in shore-to-ship transmissions—from various maritime public coast stations, AT&T's high-seas radiotelephone stations, time-signal stations, fishing and commercial fleets and other sources.

Skip signals. Shortwave is the same 3 to 30 MHz as high frequency (hf) in the electromagnetic spectrum. The extended distances over which shortwave radio signals can travel, or "propagate," depends on transmitter power, type of antenna, frequency and the condition of the atmosphere.

The ionosphere is a thick layer of Earth's upper atmosphere from 50 to 400 miles above the surface of the planet. It is described as four layers labeled D, E, F1 and F2. The ionosphere sometimes is referred to as the Kennelly-Heaviside layer, after British and American scientists who discovered its effects.

Sunspots are magnetic storms on the Sun. Their number increases to a maximum and decreases to a minimum every nine to fourteen years, averaging eleven years. This alternating rise and fall is the Sunspot Cycle. The most recent maximum was in 1990-1991.

The strange spots on the Sun affect radio transmissions on Earth. Depending on time of day and season of the year, radio waves from 500 KHz up to 25 MHz can be reflected by the ionosphere, allowing them to be received at long distances from a transmitter. These so-called sky-wave signals, bouncing from the ionosphere back down to Earth, are said to skip. Skip signals, bent or refracted by the electricity-conducting D, E, F1 and F2 layers, can be heard hundreds or thousands of miles from a transmitter.

Such sky-wave propagation is used for global telecommunications. The area between a transmitter and the point where a signal returns to Earth is the skip zone. Stations on the surface within a skip zone receive little or none of the signal passing overhead.

Propagation range. Ultraviolet light from the Sun causes the ionosphere's reflectivity by ionizing the ionosphere. Skip distance depends on the level of ionization of the ionosphere. The density of ions in the ionosphere increases and decreases with the amount of sunlight striking the top of the ionosphere. As Earth rotates on its axis, ionization above a point on the surface varies between day and night.

The amount of reflectivity varies by radio frequency. If a receiving station can't hear a transmitter on one frequency because the receiver is in a skip zone, changing to a lower or higher frequency may improve reception by moving the receiver to the end of a skip zone.

Above 25 MHz, the ionosphere usually is transparent to radio waves and doesn't reflect them. However, radiation reaching Earth from the Sun storms can change the F2 layer into a mirror for radio waves up to 50 MHz. During such times, radio signals can skip thousands of miles from a transmitter. Extremely large solar flares, which blow up on rare occasion on the Sun, provide even more spectacular skip on Earth. During these times, the muf has been known to go as high as 200 MHz.

Daytime's greater reach. Skip is strongest during daytime in the spring and fall. Starting around daybreak at lower frequencies, skip becomes more intense and spreads to higher frequencies as the hours pass. The highest frequency, which varies from day to day, is known as maximum usable frequency (muf). After noon, skip begins to fade and the muf decreases, fading out at sundown.

When skip is strongest, U.S. East Coast listeners may hear signals from Africa and Europe in the morning and from the western U.S. in the afternoon. West Coast listeners hear skip from the East and Midwest in the morning and from Central America, South America and Asia in the afternoon.

NAVAREA Warnings

The oceans between 70 degrees south and 70 degrees north latitude are divided into 12 navigation areas. Coastal stations broadcast NAVAREA warnings to ships. The information transmitted includes ships and aircraft in distress; search and rescue operations; pollution to avoid; buoy, light, fog signal and radio-navigation troubles; reefs, rocks, and shoals discovered; dangerous wrecks; towing in shipping lanes; underwater research; drifting mines; missile firings; pipe and cable-laying operations.

Navarea I North Atlantic to coasts of Greenland, England, Northern Europe, Scandinavia
Navarea II South Europe and the Atlantic coast of Africa to 6 degrees south latitude
Navarea III Mediterranean Sea
Navarea IV Caribbean Sea & Atlantic Ocean off N. America south to 7 degrees north latitude
Navarea V Atlantic off S. America to mid-ocean 7 deg north to 35 deg 50 min south latitude
Navarea VI South Atlantic Ocean off South America
Navarea VII Ocean area around southern Africa
Navarea VIII Northern area of Indian Ocean
Navarea IX Arabian Sea area
Navarea X Ocean area around Australia
Navarea XI Western Pacific islands north of Australia from equator to 45 deg north latitude
Navarea XII Pacific off North, Central, South America south to 3 deg 25 min south latitude
Navarea XIII North Pacific Ocean off the coasts of Russia and Asia
Navarea XIV Ocean east of Australia between NAVAREAs X, XV and XVI
Navarea XV Pacific off S. America, south from 18 deg south lat, east frm 120 deg west long
Navarea XVI West of mid S. American coast, north of area XV to 3 deg 25 min south lat

Latitude is the north-south position on the globe, zero to ninety degrees from the equator. High latitudes are more than 60 degrees north or south of the equator. Low latitudes are less than 30 degrees north or south of the equator. Longitude is the east-west position on the globe, from zero to 180 degrees along the equator from the prime meridian.

Ground wave. A radio transmitting antenna blankets its local area with a ground-wave signal. Ground-wave signals travel along Earth's surface, but are absorbed rapidly and covered by other signals. Such ground-wave signals are useful for short range communication in the medium-frequency (mf) band around 2 MHz.

Short skip. So-called short skip occurs from time to time when propagation carries an hf signal up to several hundred miles farther out, beyond ground wave.

Ducting. Signals on frequencies above shortwave—such as 30-300 MHz in the very high frequency (vhf) and 300-3,000 MHz in the ultra high frequency (uhf) regions of the electromagnetic spectrum—are limited to line-of-sight propagation.

Ducting is an entirely different kind of skip affecting vhf and uhf signals. It has little or nothing to do with ionization of the ionosphere. Ducted signals are conducted along paths, or cracks in the atmosphere, between different temperature layers where they meet in the lower atmosphere below the ionosphere.

While vhf and uhf signals usually are limited to line of sight distances on Earth surface, ducting can shoot them out hundreds of miles. Ducting occurs most often in springtime near sea coasts.

Public Coast And AT&T High-Seas Stations

AT&T's three high-seas radiotelephone stations in the United States provide long-range, worldwide, two-way voice communication between ships at sea and aircraft, and telephones on land, sea or in the air. The coastal stations operate 24 hours a day, seven days a week, serving commercial shipping, cruise ships, fishing boats, pleasure craft and private aircraft. They also broadcast weather forecasts and data.

The lists below show station callsign, state and town location, modulation mode, coordinated universal time (utc), and coastal station transmit frequency in MHz. Most broadcasts are in easily-copied single-sideband (ssb) voice mode. Data mode is SITOR/DSC data broadcasts. The data frequency shown is center frequency.

Weather Broadcasts By AT&T High-Seas Radiotelephone Stations

Station	Location	Mode	UTC	MHz
KMI	California, Inverness	voice	0000z	4.402
KMI	California, Inverness	voice	0000z	13.083
KMI	California, Inverness	voice	1200z	4.402
KMI	California, Inverness	voice	1200z	13.083
KMI	California, Inverness	data	odd hrs :20	8.087
WOM	Florida, Fort Lauderdale	voice	1300z	4.363
WOM	Florida, Fort Lauderdale	voice	1300z	8.722
WOM	Florida, Fort Lauderdale	voice	1300z	13.092
WOM	Florida, Fort Lauderdale	voice	1300z	17.242
WOM	Florida, Fort Lauderdale	voice	1300z	22.738
WOM	Florida, Fort Lauderdale	voice	2300z	4.363
WOM	Florida, Fort Lauderdale	voice	2300z	8.722
WOM	Florida, Fort Lauderdale	voice	2300z	13.092
WOM	Florida, Fort Lauderdale	voice	2300z	17.242
WOM	Florida, Fort Lauderdale	voice	2300z	22.738
WOO	New Jersey, Manahawkin	voice	1200z	4.387
WOO	New Jersey, Manahawkin	voice	1200z	8.749
WOO	New Jersey, Manahawkin	voice	2200z	4.387
WOO	New Jersey, Manahawkin	voice	2200z	8.749
WOO	New Jersey, Manahawkin	data	even hrs :20	8.0515

AT&T High-Seas Radiotelephone Station Addresses

AT&T Station KMI
P.O. Box 9
Inverness, CA 94937
(415) 669-1055

AT&T Station WOM
1340 N.W. 40th Avenue
Fort Lauderdale, FL 33313
(305) 587-0910

AT&T Station WOO
P.O. Box 550
End of Beach Avenue
Manahawkin, NJ 08050
(609) 597-2201

AT&T High-Seas Operator
(800) SEA-CALL

Shortwave Weather Broadcasts By Maritime Public Coast Stations

Station	Location	Mode	MHz
Alabama, Mobile	WLO	voice	2.572
California, San Francisco	KLH	voice	2.450, 2.506
Florida, Jacksonville	WNJ	voice	2.566
Florida, Miami	WDR	voice	2.442, 2.490, 2.514
Florida, Tampa	WFA	voice	2.466, 2.550
Louisiana, New Orleans	WAK	voice	2.482, 2.598, 4.419
Maryland, Baltimore	WMH	voice	2.400
Massachusetts, Boston	WOU	voice	2.450, 2.506, 2.566
New York, New York	WQX	voice	2.482, 2.522, 2.590
Texas, Corpus Christi	KCC	voice	2.538
Texas, Galveston	KQP	voice	2.530
Virginia, Norfolk	WAE	voice	2.450, 2.538
Virginia, Norfolk	WGB	voice	2.450, 2.538
Washington, Seattle	KOW	voice	2.482, 2.522

Iceberg Patrols

The cold Labrador Current carries icebergs from glaciers of western Greenland south through the Atlantic Ocean to waters off the Grand Banks where the 70-degree-warmer Gulf Stream melts the icebergs, half of the time creating dense clouds of fog.

The Grand Banks are one of the more dangerous maritime travel regions on the globe. Icebergs, fogs and powerful North Atlantic storms threaten the heavy trans-Atlantic shipping, fleets of fishing vessels and oil platform rigs. Sightings of icebergs by ships off the Grand Banks are supplemented by patrolling U.S. Coast Guard surveillance flights and other aircraft. Findings are reported to U.S. and Canadian Coast Guard stations.

The list below shows selected ice patrol broadcast station callsigns, station locations, broadcast times, modes and transmit frequencies in MHz. USCG comm stn is U.S. Coast Guard communications station. CW is continuous wave International Morse code. Fax is radiofacsimile. Fax frequency is ± 400 Hz.

Station	Location	Time	Mode	MHz
NIK	USCG comm stn Boston	0018z	voice	5.320
NIK	USCG comm stn Boston	0018z	voice	8.502
NIK	USCG comm stn Boston	0018z	voice	12.750
NIK	USCG comm stn Boston	0050z	cw	5.320
NIK	USCG comm stn Boston	0050z	cw	8.502
NIK	USCG comm stn Boston	0050z	cw	12.750
NIK	USCG comm stn Boston	0445z	navtex	0.518
NIK	USCG comm stn Boston	1045z	navtex	0.518
NIK	USCG comm stn Boston	1218z	voice	5.320
NIK	USCG comm stn Boston	1218z	voice	8.502
NIK	USCG comm stn Boston	1218z	voice	12.750
NIK	USCG comm stn Boston	1250z	cw	5.320
NIK	USCG comm stn Boston	1250z	cw	8.502
NIK	USCG comm stn Boston	1250z	cw	12.750
NIK	USCG comm stn Boston	1600z	fax	8.502
NIK	USCG comm stn Boston	1600z	fax	12.750
NIK	USCG comm stn Boston	1645z	navtex	0.518
NIK	USCG comm stn Boston	2245z	navtex	0.518
NAM/NAR/NMN/NRK:				
NAM	USCG comm stn Norfolk	0800-0900z	voice	8.090
NAM	USCG comm stn Norfolk	0800-0900z	voice	12.135

NAM	USCG comm stn Norfolk	0800-0900z	voice	16.180
NAM	USCG comm stn Norfolk	0800-0900z	voice	20.225
NAM	USCG comm stn Norfolk	1500-1600z	voice	8.090
NAM	USCG comm stn Norfolk	1500-1600z	voice	12.135
NAM	USCG comm stn Norfolk	1500-1600z	voice	16.180
NAM	USCG comm stn Norfolk	1500-1600z	voice	20.225
NAM	USCG comm stn Norfolk	1600-1700z	voice	8.090
NAM	USCG comm stn Norfolk	1600-1700z	voice	12.135
NAM	USCG comm stn Norfolk	1600-1700z	voice	16.180
NAM	USCG comm stn Norfolk	1600-1700z	voice	20.225
NAM	USCG comm stn Norfolk	2100-2200z	voice	8.090
NAM	USCG comm stn Norfolk	2100-2200z	voice	12.135
NAM	USCG comm stn Norfolk	2100-2200z	voice	16.180
NAM	USCG comm stn Norfolk	2100-2200z	voice	20.225
	USCG iceberg patrol airplane-to-ground	as needed	voice	5.696
	USCG iceberg patrol airplane-to-ground	as needed	voice	8.984
	USCG iceberg patrol helicopter-to-ground	as needed	voice	5.692
	USCG iceberg patrol helicopter-to-ground	as needed	voice	8.984

NAVTEX

NAVTEX is a global maritime bulletin service, transmitting to teleprinters worldwide on the frequency of 518 KHz. Data transmission mode is SITOR (TOR Mode B—FEC) with a shift of 170 Hz. Some American NAVTEX stations are on the air, with plans for additional NAVTEX stations on the West Coast. Canada is planning a station at Halifax, Nova Scotia. Here is a selection of stations from the Navtext schedule:

Location	Call	Area	Times
Boston, Massachusetts	NMF	F	0445, 1045, 1645, 2245
Portsmouth, Virginia	NMN	N	0130, 0730, 1330, 1930
Miami, Florida	NMA	A	0000, 0600, 1200, 1800
New Orleans, Louisiana	NMG	G	0300, 0900, 1500, 2100
San Juan, Puerto Rico	NMR	R	0415, 1015, 1615, 2215

Weather From Time Signal Stations

Various nations operate stations transmitting standard time announcements. Some of these stations also relay warnings of major weather events, particularly on the high seas. Some transmit reports on Earth's atmopshere and solar events.

Best known is WWV, operated at Fort Collins, Colorado, by the U.S. government's National Institute of Standards and Technology. It transmits 24 hours a day on the shortwave frequencies of 2.5, 5, 10, 15 and 20 MHz. Its sister station WWVH transmits continuously from Kauai, Hawaii, on 2.5, 5, 10, and 15 MHz.

The two stations broadcast time announcements every minute. Hourly, they transmit reports on weather, solar conditions, and the status of U.S. Navstar global positioning system (GPS) satellites. A brief description of Navstar is broadcast by WWV at 14 minutes past each hour. The GPS update is broadcast by WWV at 15 minutes past each hour. The Navstar introduction is broadcast by WWVH at 43 minutes past each hour and the update report is broadcast at 44 minutes past. Navstar information broadcasts are prepared by the U.S. Coast Guard's Omega Navigation Systems Center (ONSCEN) in Alexandria, Virginia, and updated once a day. Omega status reports also are announced. WWV and WWVH transmit solar activity and geomagnetic field reports.

Weather and status reports are read between time announcements. The hourly content of WWV and WWVH is shown below. The numbers represent minutes past the hour.

WWV	**WWVH**
:08 North Atlantic Ocean weather	:43 GPS introduction
:09 North Atlantic Ocean weather continued	:44 GPS status report
:10 East Pacific Ocean weather	:45 Solar activity/geomagnetic field report
:14 GPS introduction	:47 Omega status report
:15 GPS status report	:48 Pacific Ocean weather
:16 Omega status report	:49 Pacific Ocean weather continued
:18 Solar activity/geomagnetic field report	:50 Pacific Ocean weather continued
	:51 Pacific Ocean weather continued

Worldwide Standard Time Stations

MHz	Station
0.050	OMA, Prague, Czechoslovakia, operates 24 hours
0.050	RTZ, Irkutsk, Russia, operates 0100-2400 utc
0.060	MSF, Rugby, Great Britain, operates 24 hours, except 1000-1400 first Tuesday
0.060	WWVB, Fort Collins, Colorado, standard time, operates 24 hours, data
0.06666	RBU, Moscow, Russia, operates 24 hours
0.075	HBG, Prangins, Switzerland, operates 24 hours
0.0775	DCF77, Mainflingen, Germany, operates 24 hours
0.162	Allouis, France, operates 24 hours except Tuesdays 0100-0500 utc
0.198	RW-166, Irkutsk, Russia, operates 2200-2100 utc
0.272	RW-76, Novosibirsk, Russia, operates 24 hours
0.418	ZSC, Capetown, South Africa, 5-minute transmission at 0755, 1655 utc
0.434	VWC, Calcutta, India, 5-minute transmission at 0825 and 1625 utc
0.435	PPR, Rio de Janeiro, Brazil, 5-min transmissions at 0125, 1425, 2125 utc
0.482	4PB, Colombo, Sri Lanka, operates 0553-0600, 1323-1330 utc
0.500	PPR, Rio de Janeiro, Brazil, 5-min transmissions at 0125, 1425, 2125 utc
0.500	VPS, Kowloon, Hong Kong, operates every even hour
1.510	HD210a, Guayaquil, Ecuador, operates 24 hours
2.500	JJY, Tokyo, Japan, standard time, operates 24 hours, am
2.500	RCH, Tashkent, Uzbekistan, operates 0500-0400 utc
2.500	WWV, Fort Collins, Colo., standard time, weather, navigation, solar rpts, 24 hrs, am
2.500	WWVH, Kauai, Hawaii, standard time, weather, navigation, solar rpts, 24 hrs, am
3.330	CHU, Ottawa, Ontario, Canada, standard time signals, operates 24 hours, am
3.810	HD210a, Guayaquil, Ecuador, utc operating hours: 0500-1700
4.2325	VPS8, Kowloon, Hong Kong, operates every odd hour 1100-2100 utc
4.244	PPR, Rio de Janeiro, Brazil, 5-min transmissions at 0125, 1425, 2125 utc
4.2475	PPR, Rio de Janeiro, Brazil, 5-min transmissions at 0125, 1425, 2125 utc
4.286	VWC, Calcutta, India, 5-minute transmission at 1625 utc
4.291	ZSC, Capetown, South Africa, 5-minute transmission at 0755, 1655 utc
4.298	CBV, Valparaiso, Chile, 5-min at 1155,1555,1955,0055 utc, 1 hr earlier Oct 15-Mar 15
4.445	NPO, Subic Bay, Phillippines, 5-minute transmission at 0555, 1155, 1755, 2355 utc
4.996	RWM, Moscow, Russia, operates 24 hours
5.000	ATA, New Delhi, India, operates 1230-0330 utc
5.000	BPM, Xian, China, utc operating hours: 1400-2400
5.000	BSF, Chung-Li, Taiwan, operates 24 hours
5.000	HD210a, Guayaquil, Ecuador, utc operating hours: 1700-1800
5.000	HLA, Taejon, Korea, operates 24 hours
5.000	IAM, Rome, Italy, utc operating hours: 0730-0830, 1030-1100, 1 hr earlier in summer
5.000	IBF, Turin, Italy, 15-min at :45 past 0600-0800 and 1000-1700 utc, hr earlier summer
5.000	JJY, Tokyo, Japan, standard time, operates 24 hours, am
5.000	LOL, Buenos Aires, Argentina, utc operating hours: 11-12, 14-15, 17-18, 20-21, 23-24
5.000	RCH, Tashkent, Uzbekistan, operates 1400-0400 utc

5.000	VNG, Canberra, Australia, operates 24 hours, am
5.000	WWV, Fort Collins, Colo., standard time, weather, navigation, solar rpts, 24 hrs, am
5.000	WWVH, Kauai, Hawaii, standard time, weather, navigation, solar rpts, 24 hrs, am
5.000	YVTO, Caracas, Venezuela, operates 24 hours
5.004	RID, Irkutsk, Russia, operates 24 hours
5.430	BPM, Xian, China, utc operating hours: every 2 hours 1000-1800
6.840	EBC, San Fernando, Spain, operates 1029-1055 utc
7.335	CHU, Ottawa, Ontario, Canada, standard time signals, operates 24 hours, am
7.600	HD210a, Guayaquil, Ecuador, utc operating hours: 1800-0500
8.000	JJY, Tokyo, Japan, standard time, operates 24 hours, am
8.461	ZSC, Capetown, South Africa, 5-minute transmission at 0755, 1655 utc
8.473	4PB, Colombo, Sri Lanka, operates 0553-0600, 1323-1330 utc
8.492	PPR, Rio de Janeiro, Brazil, 5-min transmissions at 0125, 1425, 2125 utc
8.539	VPS35, Kowloon, Hong Kong, operates every odd hour
8.542	PKX, Jakarta, Indonesia, operates 0045-0100 utc
8.634	PPR, Rio de Janeiro, Brazil, 5-min transmissions at 0125, 1425, 2125 utc
8.650	OBC3, Callao, Peru, 5-minute transmission at 1555, 2055, 0155 utc
8.677	CBV, Valparaiso, Chile, 5-min at 1155,1555,1955,0055 utc, 1 hr earlier Oct 15-Mar 15
8.721	PPE, Rio de Janeiro, Brazil, 5-min transm'n at 0025, 1125, 1325, 1825, 2025, 2325z
9.351	BPM, Xian, China, utc operating hours: 0600 and every hour 1100-2300
9.996	RWM, Moscow, Russia, operates 24 hours
10.000	ATA, New Delhi, India, operates 24 hours
10.000	BPM, Xian, China, operates 24 hours
10.000	JJY, Tokyo, Japan, standard time, operates 24 hours, am
10.000	LOL, Buenos Aires, Argentina, utc operating hours: 11-12, 14-15, 17-18, 20-21, 23-24
10.000	RCH, Tashkent, Uzbekistan, operates 0500-1330 utc
10.000	RTA, Novosibirsk, Russia, operates 0200-0500, 1400-1730, 1800-0130 utc
10.000	VNG, Canberra, Australia, operates 2200-0700z, am
10.000	WWV, Fort Collins, Colo., standard time, weather, navigation, solar rpts, 24 hrs, am
10.000	WWVH, Kauai, Hawaii, standard time, weather, navigation, solar rpts, 24 hrs, am
10.004	RID, Irkutsk, Russia, operates 24 hours
10.0204	VPS60, Kowloon, Hong Kong, operates every odd hour 0100-1500 utc
10.4405	NPO, Subic Bay, Phillippines, 5-minute transmission at 0555, 1155, 1755, 2355 utc
11.440	PLC, Jakarta, Indonesia, operates 0045-0100 utc
12.008	EBC, San Fernando, Spain, operates 0959-1025 utc
12.307	OBC3, Callao, Peru, 5-minute transmission at 1555, 2055, 0155 utc
12.687	PPR, Rio de Janeiro, Brazil, 5-min transmissions at 0125, 1425, 2125 utc
12.724	ZSC, Capetown, South Africa, 5-minute transmission at 0755, 1655 utc
12.738	PPR, Rio de Janeiro, Brazil, 5-min transmissions at 0125, 1425, 2125 utc
12.745	VWC, Calcutta, India, 5-minute transmission at 0825 utc
12.804	NPO, Subic Bay, Phillippines, 5-minute transmission at 0555, 1155, 1755, 2355 utc
14.670	CHU, Ottawa, Ontario, Canada, standard time signals, operates 24 hours, am
14.996	RWM, Moscow, Russia, operates 24 hours
15.000	ATA, New Delhi, India, operates 0330-1230 utc
15.000	BPM, Xian, China, utc operating hours: 0000-1400
15.000	BSF, Chung-Li, Taiwan, operates 24 hours
15.000	JJY, Tokyo, Japan, standard time, operates 24 hours, am
15.000	LOL, Buenos Aires, Argentina, utc operating hours: 11-12, 14-15, 17-18, 20-21, 23-24
15.000	RTA, Novosibirsk, Russia, operates 0630-0930, 1000-1330 utc
15.000	VNG, Canberra, Australia, operates 2200-0700z, am
15.000	WWV, Fort Collins, Colo., standard time, weather, navigation, solar rpts, 24 hrs, am
15.000	WWVH, Kauai, Hawaii, standard time, weather, navigation, solar rpts, 24 hrs, am
15.004	RID, Irkutsk, Russia, operates 24 hours
17.018	ZSC, Capetown, South Africa, 5-minute transmission at 0755, 1655 utc
17.096	VPS80, Kowloon, Hong Kong, operates every odd hour 2100-1300 utc

17.1944	PPR, Rio de Janeiro, Brazil, 5-min transmissions at 0125, 1425, 2125 utc
20.000	WWV, Fort Collins, Colo., standard time, weather, navigation, solar rpts, 24 hrs, am
20.500	UQC3, Khabarovsk, Russia, operates 0036-0117, 0636-0717, 1736-1817 utc
20.500	UTR3, Gorki, Russia, operates 0536-0617, 1336-1417, 1836-1917 utc
22.352	PPR, Rio de Janeiro, Brazil, 5-min transmissions at 0125, 1425, 2125 utc
22.420	PPR, Rio de Janeiro, Brazil, 5-min transmissions at 0125, 1425, 2125 utc
22.455	ZSC, Capetown, South Africa, 5-minute transmission at 0755, 1655 utc
22.536	VPS22, Kowloon, Hong Kong, operates every odd hour 0100-0900 utc
23.000	UQC3, Khabarovsk, Russia, operates 0036-0117, 0636-0717, 1736-1817 utc
23.000	UTR3, Gorki, Russia, operates 0536-0617, 1336-1417, 1836-1917 utc
25.000	UQC3, Khabarovsk, Russia, operates 0036-0117, 0636-0717, 1736-1817 utc
25.000	UTR3, Gorki, Russia, operates 0536-0617, 1336-1417, 1836-1917 utc
25.100	UQC3, Khabarovsk, Russia, operates 0036-0117, 0636-0717, 1736-1817 utc
25.100	UTR3, Gorki, Russia, operates 0536-0617, 1336-1417, 1836-1917 utc
25.500	UQC3, Khabarovsk, Russia, operates 0036-0117, 0636-0717, 1736-1817 utc
25.500	UTR3, Gorki, Russia, operates 0536-0617, 1336-1417, 1836-1917 utc

Satellite time signals. The National Institute of Standards and Technology (former National Bureau of Standards) has used two GOES satellites to relay time signals. The western satellite at 135 degrees West longitude has transmitted time signals on 468.825 MHz. The eastern satellite at 105 degrees West has transmitted on 468.8375. There has been reception interference since the frequencies are shared with the land mobile service.

Russian Fishing Fleets Weather

Russia's northern fleet of stern-trawling fish-factory ships, tankers, freezer carriers and salvage tugs follow the supply of mackerel—working the North Atlantic Ocean off the U.S. and Canadian coasts from February to May. Small trawlers, known as MBs, scout ahead of the main fleet, radioing weather and sea data to fleet headquarters at Murmansk.

The fleet observations are compiled into marine and aviation weather forecasts by the Russian Hydromet Service. They are transmitted to vessels at sea via coastal maritime stations and fed to the World Meteorlogical Organization (WMO) which synchronizes data from weather services in nearly 200 nations. A Russian mackerel fleet traweling off North Africa often is the first to spot hurricanes forming off the Cape Verde Islands.

The western North Atlantic fishing fleet usually reports to Murmansk in radioteletype (rtty) on the maritime shortwave bands at 6, 8, 12, and 16 MHz between 2200-0200 UTC. To copy rtty, an FSK-demodulator modem and personal computer are used in conjunction with a shortwave receiver. The transmissions are at 50 baud at 170 Hz shift.

Fleet communications usually are in Russian, but sometimes in English. All messages addressed to recipients in countries other than Russia are in English. Frequencies in MHz:

4.2025–4.207 MHz	16.785–16.804 MHz
6.3005–6.3115 MHz	18.893–18.898 MHz
8.3965–8.4145 MHz	22.352–22.374 MHz
12.560–12.5765 MHz	25.193–25.208 MHz

Legal Reminder

Two-way point-to-point communications are different from one-way broadcasts intended for dissemination to the general public. Privacy of point-to-point communication is protected by law. Listeners are forbidden to reveal or repeat to anyone the content of any two-way radio communication, except amateur radio communications. The prohibition includes reporting to news media.

Land Mobile Radio And TIS

Among the many Federal Communications Commission allocations for its Private Land Mobile Radio Service is a group of vhf high-band (150 to 174 MHz) channels set aside in the lower 48 states for meteorological and hydrological usage. The six land mobile user groups permitted to make use of the frequencies are:

Power (IW) Generation, transmission or distribution of electrical energy for use by the public or members of a cooperative organization. Distribution of manufactured or natural gas by pipeline, for use by the public or members of a cooperative organization, or in a combination of that with production, transmission or storage of manufactured or natural gas for distribution. Distribution of steam or water by pipeline, canal or open ditch for use by the public or members of a cooperative organization, or in a combination of that with collection, transmission, storage or purification of water or the generation of steam for distribution.

Petroleum (IP) Use in prospecting for, producing, collecting, refining or transportation by pipeline of petroleum or petroleum products including natural gas. Persons containing or cleaning up oil spills use the frequencies with limitations.

Forest Products (IF) Tree logging, tree farming, or related woods activities including hauling. Manufacturing lumber, plywood, hardboard, pulp and paper products from wood fiber.

Special Industrial (IS) Operation of farms, ranches or similar land areas for the quantity production of crops or plants; vines or trees (except for forestry operations); or for keeping, grazing or feeding of livestock for animal products, animal increase or value enhancement. Agricultural plowing, soil conditioning, seeding fertilizing or harvesting. Spraying or dusting of insecticides, herbicides or fungicides in areas other than enclosed structures. Livestock breeding service.

Commercial business constructing roads, bridges, sewer systems, pipelines, airfields or water, oil, gas or power production, collection or distribution systems, except for building construction. Mining solid fuels, minerals, metal, rock, sand and gravel from land or sea. Maintaining, patrolling or repairing gas or liquid transmission pipelines, tank cars, water or waste disposal wells, industrial storage tanks or distribution of public utilities. Acidizing, cementing, logging, perforating, shooting or similar activities in drilling new oil or gas wells or maintenance of production from established wells.

Supplying chemicals, mud, tools, pipe and equipment unique to the petroleum and gas production industry, if the application of materials needs special conveyances. Delivery of ice or fuel to a consumer for heating, lighting, refrigeration or power by means other than pipelines or railroads when such are not to be resold following delivery. The delivery and pouring of ready mixed concrete or hot asphalt mix.

Business (IB) Operation of a commercial activity. Operation of an educational, philanthropic, or ecclesiastical institution. Activities of clergymen. Operation of hospitals, clinics or medical associations.

Railroad (LR) Railroad common carriers regularly engaged in transportation of passengers or property transported over all or part of their route by railroad.

MHz	Users
169.4250	meteorological use by IW, IP, IF, IS
169.4500	meteorological use by IW, IP, IF, IS, IB, LR
169.4750	meteorological use by IW, IP, IF, IS, IB, LR

169.5000	meteorological use by IW, IP, IF, IS, IB, LR
169.5250	meteorological use by IW, IP, IF, IS, IB, LR
170.2250	meteorological use by IW, IP, IF, IS, IB, LR
170.2500	meteorological use by IW, IP, IF, IS, IB, LR
170.2750	meteorological use by IW, IP, IF, IS, IB, LR
170.3000	meteorological use by IW, IP, IF, IS, IB, LR
170.3250	meteorological use by IW, IP, IF, IS, IB, LR
171.0250	meteorological use by IW, IP, IF, IS, IB, LR
171.0500	meteorological use by IW, IP, IF, IS, IB, LR
171.0750	meteorological use by IW, IP, IF, IS, IB, LR
171.1000	meteorological use by IW, IP, IF, IS, IB, LR
171.1250	meteorological use by IW, IP, IF, IS, IB, LR
171.8250	meteorological use by IW, IP, IF, IS, IB, LR
171.8500	meteorological use by IW, IP, IF, IS, IB, LR
171.8750	meteorological use by IW, IP, IF, IS, IB, LR
171.9000	meteorological use by IW, IP, IF, IS, IB, LR
171.9250	meteorological use by IW, IP, IF, IS, IB, LR

IW	Power
IP	Petroleum
IF	Forest Products
IS	Special Industrial
IB	Business
LR	Railroad

Traveler's Information Service

For many years, the medium-wave a.m. broadcast band in the U.S. has covered 540 to 1600 KHz. The upper limit will be extended another 100 KHz in the 1990s, but for now there is a special collection of low-power a.m. stations just above and below the band where weather forecasts sometimes can be found.

Known as Traveler's Information Service (TIS), most of the stations are on 530 and 1610 KHz. TIS stations in the U.S. and Canada repeat tape-recorded public-service announcements near highways, airports, parks, and tourist attractions. In the future, TIS stations will be spread throughout the medium-wave a.m. broadcast band.

TIS geographic coverage is very limited as the transmitters put out only about five watts of power to a short whip antenna. However, sometimes propagation conditions allow one of these medium-wave pip-squeaks to be heard as much as 500 miles from its antenna—when no closer station interferes. Unfortunately for DXers trying to hear Traveler's Information Service stations on 1610 KHz, a 50 kilowatt religious broadcaster on Anguilla in the British West Indies often blocks out distant TIS signals. To further complicate TIS listening, illegal pirate broadcasters keep popping up on 1610 KHz.

Weather. A handful of TIS stations are wired to relay forecasts originated by local NOAA Weather Radio stations found on VHF narrow-band fm (nbfm) around 162 MHz. One example is a Hagerstown, Maryland, TIS station on 530 KHz.

Service	Frequency
Traveler's Information Service	530 KHz
Traveler's Information Service	1610 KHz

Amateur Radio In Weather Crises

Amateur Radio is a non-commercial radio communication service permitting communication between private persons—usually referred to as hams—for hobby purposes and for public-service. Hams have fun chatting with their neighbors, with friends hundreds of miles away and with amateur radio operators on other continents.

Hams pass relatively-easy tests to obtain their federal licenses. There are some 500,000 amateur radio operators in the USA alone and more than a million elsewhere around the globe. Male and female hams range in age from 5 to 95. Many physically-challenged persons enjoy amateur radio.

Hams use their privileges to chat with friends, near and far, at home or while walking, driving, boating and flying; converse with people in foreign countries; help with communications in emergencies; provide communications for parades, bike-a-thons and walk-a-thons; and teach others to be hams. Radio amateurs serve local civil defense operations through the Amateur Radio Emergency Service (ARES) and the Radio Amateur Civil Emergency Service (RACES). They also participate in contests and Field Days.

Amateur radio provides someone to talk with during sleepless nights and can teach electronics. Hams talk with astronauts in space, bounce signals off the Moon and receive weather pictures from satellites.

Amateur operators communicate by voice, International Morse Code, radioteletype, slow-scan and fast-scan television, fax and computer. They collect confirmations of their contacts with others in the form of QSL cards from around the globe. QSL cards qualify hams for achievement awards.

Amateurs, however, are not allowed to transact business by ham radio. They also are not allowed to interfere with other hams and other radio services. Transmitting music, obscene, profane or indecent language, and one-way broadcasting is illegal.

Thousands of hams volunteer for emergency service, practicing regularly on shortwave, VHF and UHF bands. Most any time, there are organized on-the-air networks passing weather information and simulated emergency messages, or "traffic." When a natural or man-made disaster breaks out, hams can be heard relaying weather data, health-and-welfare traffic and sometimes life-or-death messages.

The American Radio Relay League (ARRL) is the national fraternity of amateur radio operators, headquartered at 225 Main St., Newington, CT 06111.

Amateur Radio Weather Emergency Nets

There are hundreds of amateur radio public-service communications networks, ready to be activated in the event of local, state, regional, national or international weather emergency or man-made disaster.

The weather-tracking and emergency-preparedness nets in the master list below are arranged in alphabetical order under each state. Following the individual states are U.S. regions, Canada, Mexico and international. There are hundreds of other amateur nets, organized for purposes other than emergency preparedness, but they are not listed here.

The frequencies, in megahertz (MHz), of the weather nets are approximate because heavy use of a frequency, or unusual propagation conditions, sometimes make it necessary for a net to move slightly up or down the band. After all, in amateur radio, no group has a preemptory right to a specific frequency.

Amateur radio two-meter fm repeaters operate on pairs of frequencies between 144 and 148 MHz. The input frequency of a two-meter repeater usually is removed from the output by 600 KHz. For example, a ham might transmit to a repeater input on 146.13 and the repeater would repeat the signal on an output of 146.73. Repeater outputs are listed below. If nothing is heard on an output frequency, try monitoring the input frequency.

Net starting times are UTC, also known as GMT, except for the few nets marked local time. *(UTC is explained on page 7.)* Nets which meet year-round at a local time meet an hour earlier UTC when Daylight Saving Time is in effect.

The abbreviations used in the lists below and by amateurs during communications are:

Frequency bands

ulf	ultra low frequency; 30 Hz to 300 Hz
vlf	very low frequency; 300 Hz to 30 KHz; audio frequencies
lf	low frequency; 30 KHz to 3 MHz
hf	high frequency; shortwave; 3 MHz to 30 MHz
vhf	very high frequency; 30 MHz to 300 MHz
uhf	ultra high frequency; 300 MHz and above
KHz	frequency in kilohertz
MHz	frequency in megahertz
80 meters	3.5 to 3.8 MHz; amateur radio hf band; usually local and regional communications
75 meters	3.8 to 4.0 MHz; amateur radio hf band; usually local and regional communications
40 meters	7.0 to 7.3 MHz; amateur radio hf band; usually local and regional communications
20 meters	14.0 to 14.35 MHz; amateur radio hf band; usually regional and international communications
15 meters	21.0 to 21.45 MHz; amateur radio hf band; usually regional and international communications
10 meters	28.0 to 29.7 MHz; amateur radio hf band; usually regional and international communications
6 meters	50 to 54 MHz; amateur radio vhf band; usually local and regional communications
2 meters	144 to 148 MHz; amateur radio vhf band; local and regional communications
1.25 meters	222 to 225 MHz; amateur radio vhf band; usually local and regional communications
70 cm	420 to 450 MHz; amateur radio uhf band; usually local and regional communications
alt.	alternate

Days

Dy	daily	W	Wednesday
Sn	Sunday	Th	Thursday
M	Monday	F	Friday
T	Tuesday	Sa	Saturday

Modes of operation

am	amplitude modulation voice mode of communication
amtor	keyboard mode of communication similar to packet but on hf
atv	amateur television mode of communication
bbs	radio bulletin board system; a packet radio communications network
cw	continuous wave, International Morse code mode of communication
digi	digipeater; a packet radio repeater
digipeater	a packet radio repeater
fax	facsimile, a picture mode of radio communication
fm	narrow-band frequency modulation voice mode of communication
fstv	fast-scan television mode of communication
lsb	lower sideband single-sideband voice mode of communication
node	a "smart" digi or more-programmable digipeater
packet	keyboard mode of communication; connects with a bbs or communications network
pbbs	packet bulletin board system; a radio communications network
pkt	packet; keyboard mode of communication; radio bulletin board system
rbbs	radio bulletin board system; a communications network
repeater	a voice radio signal repeating bridge between two amateur stations
rptr	repeater; a voice radio signal repeating bridge between two amateur stations
rtty	radioteletype; a keyboard mode of communication
ssb	single sideband, a voice mode of communication with less bandwidth than am or fm
sstv	slow-scan television mode of communication
usb	upper sideband single-sideband voice mode of communication

Volunteer emergency groups

A/R	ARES/RACES
ARC	amateur radio club
ARES	Amateur Radio Emergency Service
CAP	Civil Air Patrol; volunteers search for downed aircraft on frequencies outside of ham bands
MARS	Military Affiliate Radio System, amateurs providing free communications for the military
net	network; several individual amateur stations meeting on a frequency at the same time
NTS	National Traffic System; amateurs sending routine and emergency radiograms for third parties
RACES	Radio Amateur Civil Emergency Service
SAREX	in amateur radio: Shuttle Amateur Radio Experiment; in CAP: Search And Rescue Exercise
SKYWARN	amateur radio net feeding severe weather data to the National Weather Service

Q codes used in Morse Code and nets

QRT	stop operating		QSL	confirmation
QRU	I have no traffic		QSO	conversation
QRV	I am ready to copy traffic		QSY	change frequency
QRX	stand by			

Miscellaneous abbreviations

Co.	county
DX	distance
Emerg.	Emergency
ht	handy talky; a hand-held, battery-powered two-way radio
Sct.	Section
UTC	Coordinated Universal Time; same as GMT, Greenwich Mean Time
wx	weather

Net	Mode	MHz	Day	UTC
Alabama				
135 Repeater Group	fm	147.135	Th	0200
Alabama ARES-RACES Net	lsb	3.965	T	0130
Alabama ARES-RACES Net	lsb	7.260	S	1630
Alabama ARES/RACES Net	lsb	3.926	T	0130
Alabama ARES/RACES Net	lsb	7.260	Sa	1630
Alabama Emergency Net G, south central	fm	147.18	M,Th	
Alabama Emergency Net I, east central	fm	147.00	W	0200
Alabama Emergency Net J, northwest Alabama	fm	146.61	Dy	0000
Alabama Emergency Net K, Montgomery	fm	146.91	Th	0130
Alabama Emergency Net L, southeast Alabama	fm	146.76	W	0100
Alabama Emergency Net N, Shelby Co.	fm	146.98	T	0200
Alabama Emergency Net O, Butler Co.	fm	146.67	Th	0100
Alabama Emergency Net P, Dale Co.	fm	146.85	F	0300
Alabama Emergency Net Q, Covington	fm	146.94	T	0100
Alabama Emergency Net R, Madison	fm	50.52	T,Th	0100
Alabama Emergency Net S, northeast Alabama	fm	147.06	W	0200
Alabama Emergency Net T, Cullman Co.	fm	146.85	Sn	0300
Alabama Emergency Net U, Tuscaloosa	fm	147.30	W	0230
Alabama Emergency Net V, east and central Alabama	fm	147.36	Th	0130
Alabama Emergency Net W, east central Alabama	fm	147.09	M	0200
Alabama Emergency Net X, central Alabama	fm	146.88	T	0300
Alabama Emergency Net Y, Etowah Co.	fm	147.16	T	0230
Alabama Emergency Net Z, DeKalb Co.	fm	147.27	T,S	0030
Baldwin County Emergency Net	fm	146.685	Sn	2100
Central Alabama 2-Meter Net	fm	146.64	M	0230
Chattanooga Area Weather Net	fm	146.61	W	0100
East Central Alabama ARES Net	fm	147.06	F	0200
Lee Co. Emergency Training Net	fm	147.12	Th	0200
Limestone ARES Net, north central Alabama	fm	147.20	F	0100
Limestone ARES Net, north central Alabama	fm	147.20	F	1930

North Alabama Severe Weather Net	fm	145.11	Dy	1900
North Georgia 220 Net, northeast Alabama	fm	224.74	on call	
South Alabama SKYWARN Net	fm	147.18	M	0215
Tri State Two Meter Net	fm	147.300	Sn,T,Th	2130
West Alabama ARES Net	fm	146.64	M	0000
West Alabama Emergency Net, Tuscaloosa	fm	146.82	M	0130

Alaska

Alaska Bush Net	cw	7.091	Dy	0500
Alaska Longwire Net		1.843	T,Th,S	0700
Alaska Pacific Net, west coast	usb	14.292	M-F	1730
Alaska Tsunami Emergency Net	usb	3.920	on call	
Anchorage ARES/RACES Net	fm	147.30	Th	0500
Juneau ARES Net	fm	147.30	S	1800
Kodiak ARES Net	fm	146.88	on call	
Tanana Valley Preparedness Net, Fairbanks	fm	146.79	M	0615
Tanana Valley Preparedness Net, Fairbanks	usb	28.400	M	0600

Arizona

Arizona State RACES	lsb	3.9905	Sn	1500
Arizona Traffic and Emergency Net (summer)	lsb	3.992	Dy	0230
Arizona Traffic and Emergency Net (winter)	lsb	3.992	Dy	0200
Coconino Co. ARES Net	fm	147.38	W	0200
London Bridge ARES Net, south Mohave Co.	fm	146.61	M	0200
Mercury Amateur Radio Association	lsb	3.983	S	1500
Mohave Co. ARES Net	fm	146.76	M	0230
Nevada Weather Net	lsb	3.993	M-Sa	0600
Prescott/Yavapai Co. ARES	fm	146.88	T,Th	0230
Tucson/Pima Co. RACES	fm	147.30	S	2000
Tucson/Pima Co. RACES	lsb	3.995	Sn	1545
West Valley 2-Meter Net, West Maricopa Co.	fm	147.30	W	0200
Yavapai Co ARES Net	fm	147.82	Th	0200
Yuma ARES What Net	fm	146.74	F	2200

Arkansas

Arkansas Emergency Communications Net	lsb	3.9875	Sn	2330
Arkansas Phone Net	lsb	3.885	M-Sa	1200
Arkansas Razorback Net	lsb	3.987.5	Dy	0030
Arkansas Weather Net	fm	146.94	on call	
Arkansas Weather Net	lsb	3.987	on call	
Baxter County Weather Net	fm	146.88	T	0200
Central AR Weather Net	fm	146.94	on call	
Central Ark. Radio Emergency Net	fm	146.94	Th	0200
East Arkansas Weather Net	fm	146.61	W	0300
Faulkner County Net	fm	146.97	Th	0130
Ft. Smith Radio Club Net	fm	146.64	Th	0200
Hot Springs County Emergency Net	fm	147.36	F	0100
Northwest Arkansas ARC Net	fm	146.76	M	0200
OARA Net, West Central Arkansas	fm	146.79	W	0200
Polk County ARES Net	fm	146.82	Th	0200
South Central AR Weather Net	fm	147.36	on call	
Southwest Missouri SKYWARN Net	fm	146.91	W	0100

California

Alameda ARES/RACES Net	fm	145.00	F	0300
Alameda Co. Emergency Net	fm	147.24	T	0300
Alameda Co. RACES Net	lsb	3.987	F	0300
Antelope Valley ARC, Mojave Desert	fm	146.73	Th	0300
Baja California Maritime Net	lsb	7.2385	Dy	1600

Net	Mode	Freq	Day	Time
Belmont/San Carlos/Redwood City ARES	fm	147.45	M	0330
Berkeley Emergency ARS	fm	146.43	Th	0300
Butte Co. Amateur Radio Emergency Service Net	fm	146.85	M	0430
California to Caribbean, maritime service net	usb	14.285	M	2300
California to South Pacific, maritime service net	usb	14.285	M	2310
California Traffic Net, western U.S.	lsb	3.905	Dy	0230
Campbell ARES/RACES	fm	145.45	ET	2245
CARDA Rescue Dog Net	fm	149.925	2nd T	2000 local
Central Contra Costa Co. ARES/RACES Net	fm	146.43	F	0300
Central San Joaquin Valley Health & Welfare Traffic Net	fm	147.33	W	0200
Contra Costa Co. ARES Net	fm	147.135	F	0320
Corona Norco Ama Radio Club RACES Net, Riverside	fm	146.535	T	0400
Cupertino ARES/RACES	fm	147.57	T	1945 local
Daly City/S. SF ARES	fm	146.505	W	0300
Eastern Contra Costa Co. ARES/RACES Net	fm	147.54	F	0300
Gilroy City ARES/RACES	fm	147.625	T	1945 local
Glenn Co. Amateur Radio Emergency Service Net	fm	146.85	M	0430
Goin' Home Group, Northern SF Bay area	fm	145.11	M-F	2100
Half Moon Bay RACES, West San Mateo Co.	fm	147.285	W	0300
Hemet Valley ARES-RACES 2-Meter Net	fm	145.42	W	0330
Hemet Valley Emergency Net	lsb	3.945	W	0400
High Desert ARES Net, north Los Angeles Co.	fm	52.16	T	0300
High Desert ARES Net, north Los Angeles Co.	fm	147.555	T	0300
High Desert ARES Net, So. Kern Co.	fm	52.16	T	0300
High Desert ARES Net, So. Kern Co.	fm	147.555	T	0300
Indian Wells Valley Emergency Net	fm	224.64	M	0330
Insomnia Net, Southern California	fm	144.33	Dy	0700
Insomnia Net, Southern California	fm	145.45	on call	
Inyo Co. Orange Sct. ARES N. Alameda Co. ARES Net	fm	146.94	Dy	0500
Inyo Co. Orange Sct. ARES N. Alameda Co. ARES Net	fm	147.48	F	0320
Inyo Co. Orange Sct. ARES N. Alameda Co. ARES Net	fm	440.90	F	0320
Jumpsuit Net	ssb	14.180	M-F	2000
La Habra Heights Disaster Communication Services	fm	146.46	T	0230
Livermore ARC RACES Net	fm	147.12	T	0300
Lompoc ARES	fm	145.50	M	0300
Los Altos ARES/RACES	fm	145.57	T	2000 local
Los Altos Hills ARES/RACES	fm	147.435	T	2000 local
Los Angeles Ama. Radio Emergency Support Team	fm	145.58	2nd/4th T	0400
Los Angeles Ama. Radio Emergency Support Team	fm	224.34	2nd/4th T	0400
Los Gatos ARES/RACES	usb	28.485	T	1945
Los Gatos ARES/RACES	fm	145.45	T	1945
Los Gatos ARES/RACES	fm	222.22	T	1945
Marin County 2 Meter Emergency Net	fm	146.70	Sn	1820
Marin Red Cross Emergency Net	lsb	3.915	Sn	1800
Menlo Park ARES	fm	147.45	T	0330
Mercury Amateur Radio Association	lsb	3.983	S	1500
Millbrae ARES	fm	146.49	Th	0330
Milpitas ARES/RACES	fm	144.135	T	1915 local
Milpitas ARES/RACES	fm	224.72	T	1915 local
Monterey ARES	fm	146.97	W	1930 local
Monterey ARES	fm	444.7	W	1930 local
Monterey ARES	fm	445.1	W	1930 local
Monterey Bay Emergency Net	fm	144.45	M	1930
Morgan Hill ARES/RACES	fm	147.285	T	1945 local
Mountain View ARES/RACES	fm	145.25	T	2000 local

Net	Mode	Freq	Days	Time
Napa Co. ARES/RACES	fm	147.18	Th	0300
Nevada Weather Net	lsb	3.993	M-Sa	0600
No. Santa Barbara Co. ARES	fm	145.14	M	0400
No. SF Bay Counties Afternoon 220 Net	fm	220.30	M-F	2200
North Orange Co. Orange Sct. ARES	fm	146.79	W	0330
Northern Alameda Co. ARES Net	fm	147.48	F	0320
Northern Alameda Co. ARES Net	fm	440.90	F	0320
Northern California Net	cw	3.630	Dy	0300
Northern California Net	fm	145.41	Dy	0330
Orange Co. ARES	lsb	3.965	Sn	1600
Orange Co. Huntington Beach Orange Sct. RACES	fm	145.14	T	0330
Orange Co. RACES	fm	52.62	T	0300
Orange Co. RACES	fm	146.895	T	0300
Orange Section Emergency Net ARC	fm	144.85	W	2000
Pacifica ARES/RACES	fm	146.926	W	2100 local
Palo Alto ARES/RACES	fm	145.27	M	2000 local
Portola Valley EC Net, San Francisco vicinity	fm	146.865	T	0300
Redwood City Disaster Services	fm	147.45	T	0230
Riverside Co. ARES/RACES #1	fm	146.88	T	0330
Riverside Co. ARES/RACES #2, Corona	fm	147.15	T	0400
Riverside Co. ARES/RACES #2, Norco	fm	147.15	T	0400
Riverside Co. ARES/RACES #3, Banning	lsb	3.987	Th	0230
Riverside Co. ARES/RACES #3, Beaumont	lsb	3.870	Th	0230
Riverside Co. ARES/RACES #4, Yucaipa	fm	147.57	T	0315
Riverside Co. ARES/RACES #5, PalmSprgs	fm	146.94	T	0300
Riverside Co. ARES/RACES #6, Coachell	fm	146.625	F	0300
Riverside Co. ARES/RACES #6, Indio	fm	146.625	F	0300
Riverside Co. ARES/RACES #7, Blythe	fm	146.85	Th	0300
Riverside Co. ARES/RACES #7, Palo Verde	fm	146.85	Th	0300
Riverside Co. ARES/RACES #8, Elsinore	fm	146.76	T	0300
Riverside Co. ARES/RACES #9	fm	147.585	1st T	0300
Riverside Co. ARES/RACES/VIP #10	lsb	3.945	Sn	1630
Riverside Co. ARES/RACES #11, Banning	fm	144.44	Th	0315
Riverside Co. ARES/RACES #11, Beaumnt	fm	144.44	Th	0315
Riverside Co. ARES/RACES #12, SunCity	fm	146.49	Th	0230
Riverside Co. ARES/RACES #13, Coachella	fm	146.70	M	0300
Riverside Co. ARES/RACES #14, Western	fm	224.46	T	0315
Riverside Co. Hq. EOC ARES/RACES	fm	145.20	T	0315
S. Peninsula Emergency Communications, San Francisco	fm	145.27	T	0400
S. Peninsula Emergency Communications, San Francisco	fm	224.36	T	0400
Sacramento Amateur Emergency Net	fm	146.91	M	2015
Sacramento City RACES/County ARES Net	fm	147.195	M	1900
Sacramento Valley SKYWARN Net	fm	147.015	M-F	1600
San Bernardino Co. ARES 2-Meter	fm	146.85	Th	0300
San Bernardino Co. ARES 10-Meter	usb	29.200	T	0200
San Bernardino Co. ARES 220 Net	fm	224.86	Th	0400
San Bernardino Co. ARES/RACES #1, W San Bernardino	fm	147.48	T	0300
San Bernardino Co. ARES/RACES #2, Fontana	fm	145.52	T	0330
San Bernardino Co. ARES/RACES #2, Rialto	fm	145.52	T	0330
San Bernardino Co. ARES/RACES #3, San Bernardino	fm	145.32	T	0300
San Bernardino Co. ARES/RACES #4, Redlands	fm	147.57	T	0315
San Bernardino Co. ARES/RACES #4, Yucaipa	fm	147.57	T	0315
San Bernardino Co. ARES/RACES #5, Big Bear	fm	146.985	1st,3rd W	0400
San Bernardino Co. ARES/RACES #5, Crestline	fm	146.985	1st,3rd W	0400
San Bernardino Co. ARES/RACES #6, Hesperia	fm	146.94	M	0400

San Bernardino Co. ARES/RACES #6, Victorville	fm	146.94	M	0400
San Bernardino Co. ARES/RACES #7, Barstow	fm	147.15	T	0230
San Bernardino Co. ARES/RACES #7, Hinkley	fm	147.15	T	0230
San Bernardino Co. ARES/RACES #8, Morongo Valley	fm	146.79	W	0300
San Bernardino Co. ARES/RACES #9, Needles	fm	146.76	T	0330
San Bernardino Co. ARES/RACES #10, Baker	fm	147.57	T	0300
San Bernardino Co. ARES/RACES #10, Mountain Pass	fm	147.57	T	0300
San Bernardino Co. Hq. EOC ARES/RACES	fm	145.20	T	0315
San Bernardino Co. Hq. EOC ARES/RACES, Colton	fm	147.45	T	0325
San Bernardino Co. Hq. EOC ARES/RACES, Redlands	fm	147.45	T	0325
San Bernardino Co. Hq. EOC ARES/RACES, Riverside	fm	145.20	T	0315
San Bernardino Co. RACES 2-Meter	fm	147.345	T	0345
San Bernardino Co. RACES 220 MHz	fm	224.86	T	0345
San Bernardino Co. RACES RTTY	rtty	145.12	T	0400
San Bernardino Co. RACES/ARES	lsb	3.9875	T	0300
San Bruno ARES	fm	145.79	M	0400
San Diego Co Traffic Net	fm	146.730	Dy	0400
San Diego Emergency Net	fm	147.555	Th	0300
San Diego Section ARES Net	cw	3.725	Sn	1730
San Diego Section ARES Net	fm	144.25	Sn	0400
San Diego Section ARES Net	fm	146.48	M	0300
San Diego Section ARES Net	fm	146.73	Sn	1630
San Diego Section ARES Net	fm	224.90	Sn	0300
San Diego Section ARES Net	lsb	3.905	Sn	1700
San Diego Section ARES Net	usb	28.375	Sn	1800
San Diego Section ARES-Southern District	usb	28.450	T	0430
San Diego Section ARES-Southern District	fm	146.55	W	0400
San Diego Section-Coronado ARES	fm	147.18	Th	0330
San Diego Section-Eastern District	fm	147.57	M,Sn	0330
San Jacinto Valley ARES-RACES 2-Meter Net	fm	145.42	W	0330
San Jose ARES	fm	145.65	on call	
San Jose ARES	fm	146.035	T	0330
San Jose ARES	fm	146.43	on call	
San Jose ARES	fm	146.475	on call	
San Jose City RACES Net	fm	146.985	Th	0300
San Jose RACES, Santa Clara Co.	fm	145.65	Th	1930 local
San Jose RACES, Santa Clara Co.	fm	146.43	Th	1930 local
San Jose RACES, Santa Clara Co.	fm	146.475	Th	1930 local
San Lorenzo ARES/RACES	fm	147.12	T	1930 local
San Lorenzo ARES/RACES	fm	440.85	on call	
San Luis Obispo County ARES/RACES Net	fm	146.80	T	0330
San Mateo ARC	fm	145.64	T	2030
San Mateo ARES	fm	146.925	T	2000 local
San Mateo City ARES/RACES	fm	145.64	TTh	1930
Santa Barbara South County ARES	fm	146.79	T	0330
Santa Clara ARES/RACES	fm	146.85	T	1930 local
Santa Clara ARES/RACES	fm	147.45	T	1930 local
Santa Cruz ARES Net	fm	146.79	M	2030
Santa Cruz Co. Traffic & Emergency Training Net	fm	146.79	T	1930 local
Santa Maria ARES Net	fm	145.14	M	0300
Santa Maria ARES Net	fm	146.94 alt	M	0300
Santa Ynez Valley ARES Net	fm	146.895	M	0245
Saratoga ARES/RACES	fm	145.505	T	0345
So. Santa Barbara Co. ARES Net	fm	146.79	M	0330
South Orange Co. ARES	fm	145.14	Th	0330

South Orange Co. ARES	fm	147.045	on call	
Southern California Emergency Health and Welfare Net	lsb	3.955	T	0400
Southern California Net	cw	3.705	Dy	0600
Southern Contra Costa Co. ARES/RACES Net	fm	145.555	F	0245
Southern Solano Co. ARES/RACES Net	fm	146.70	Dy	1700
Southern Solano Co. ARES/RACES Net	fm	147.135	T	0300
Southern Solano Co. ARES/RACES Net	fm	224.72	Dy	1700
Southwestern Division ARES, So. Calif.	ssb	1.945	Sn	1900
Stanford University ARES	fm	145.27	M	2000 local
Sulfur Mountain Rptr ARES/RACES Net, Ventura Co.	fm	146.88	W	0330
Sunnyvale ARES/RACES	fm	147.905	T	1945 local
Tuolumne Co. Emergency Net	lsb	3.912	Sn	1700
Tuolumne Co. Novice Emergency Net	cw	3.710	Sn	1530
West Contra Costa Co. ARES/RACES Net	fm	145.11	F	0300
West Kern Co. ARES Net	fm	146.91	T	0200

Colorado

Colorado ARES Net	lsb	3.928	Sn	1500
Colorado ARES Net	lsb	7.230 alt.	Sn	1500
Colorado Emergency Phone Net	lsb	3.945	Sn	1500
Colorado RACES Net	lsb	3.9905	Sn,T	0100
Colorado Weather Net	lsb	3.945	Dy	1330
District 9 ARES Net, Clear Creek Co.	fm	145.37	W	0230
District 10 ARES and Information Net, Jackson Co.	fm	145.115	T	0400
District 10 ARES and Information Net, Larimer Co.	fm	145.115	T	0400
District 10 ARES and Information Net, Weld Co.	fm	145.115	T	0400
District 11 ARES Net, Boulder Co.	fm	147.27	Sn	1610
District 11 ARES Net, Boulder Cos.	fm	146.76	M	0400
District 11 ARES Net, Gilpin Co.	fm	147.27	Sn	1610
District 11 ARES Net, Gilpin Cos.	fm	146.76	M	0400
District 12 ARES Net, Morgan Co.	fm	147.24	M	0500
District 12 ARES Net, Washington Co.	fm	147.24	M	0500
District 13 ARES Net, Denver	fm	147.33	Sn	0400
District 14 ARES Net, El Paso Co.	fm	146.97	T	0200
District 14 ARES Net, Teller Co.	fm	146.97	T	0200
District 15 ARES Net, Custer Co.	fm	147.21	M	0130
District 15 ARES Net, Fremont Co.	fm	147.21	M	0130
District 16 ARES Net, Huerfano Co.	fm	146.79	Sn	0200
District 16 ARES Net, Pueblo Co.	fm	146.79	Sn	0200
District 17 ARES Net, Cheyenne Co.	fm	147.06	M	0200
District 17 ARES Net, Kit Carson Co	fm	147.06	M	0200
District 17 ARES Net, Lincoln Co.	fm	147.06	M	0200
District 21 ARES Net, Logan Co.	fm	146.91	Th	0300
District 21 ARES Net, Phillips Co.	fm	146.91	Th	0300
District 21 ARES Net, Sedgwick Co.	fm	146.91	Th	0300
District 22 ARES Net, Arapahoe Co.	fm	146.64	Sn	1600
District 23 ARES Net, Jefferson Co.	fm	146.67	Sn	1600
District 24 ARES Net, Douglas Co.	fm	147.225	Th	0300
District 27 ARES Net, Adams Co.	fm	145.43	M	0300
Eastern Slope ARES Information Net	fm	146.94	Sn	1530
Eastern Slope ARES Net	fm	146.97 alt.	Sn	1530
Estes Park ARES Net	fm	146.685	4th S	1430
Fort Collins ARES Net	fm	147.36	S	1530
Greeley ARES Net	fm	147.00	F	0400
Loveland ARES Net	fm	147.195	S	1500
Mesa Co. ARES Information	fm	146.94	S	0200

Ski Country Amateur Radio Club, western slope	fm	146.88	Sn	0400

Connecticut

Amateur Radio Emergency Service Net Area II	fm	145.29	2nd M	0030
BEARS Net	fm	224.32	T	0030
Bethel ARES Net	fm	147.45	2nd T	0030
Bethel ARES Net	fm	224.32	2nd T	0030
Bethel ARES Net	usb	28.350	2nd T	0000
Bloomfield Amateur Radio Emerg Net & Service	fm	146.82	W	0100
Connecticut Civil Preparedness-Area 1, SW Connecticut	fm	146.835	2nd T	0100
ECARA Emergency Net	fm	147.225	F	0130
Enfield ARES Net	usb	28.400	3rd Sn	0000
Larkfield Emergency Preparedness Net, south Connecticut	fm	147.21	Th	0030
Larkfield Mariners Net, south Connecticut	fm	147.21	T	0100
Milford Civil Preparedness Net	fm	146.925	T	0100
Northwest Connecticut ARES Net	fm	146.955	Th	0100

Delaware

Delaware Emergency Phone Net	lsb	3.905	S	2300
Delaware Traffic Net	lsb	3.905	M-F	1830
Delaware Traffic Net	lsb	3.905	M-F	2330
Maryland Emergency Phone Net	lsb	3.920	Dy	2300
Sussex Co. Emergency Net	fm	147.075	T	0001

Florida

Alachua Co. Emergency Net	fm	146.82	Th	0100
All Florida ARES Net	lsb	3.950	M-F	1300
Beaches ARES Corp Net, Duval Co.	fm	145.35	T	0030
Broward Emergency Net	fm	146.91	M	0000
Capital District ARES Net	fm	147.03	S	0000
Clay County Amateur Radio Emergency Service Net	fm	146.67	M	0030
Dade Emergency Net, Miami	fm	147.00	Th	0100
Dade Emergency Net, Miami	fm	224.10	Th	0100
Dade Emergency Net, Miami	fm	444.725	Th	0100
East Orange Emergency Net, Brevard	fm	147.06	S	0100
East Orange Emergency Net, Osceola	fm	147.06	S	0100
East Orange Emergency Net, Seminole	fm	147.06	S	0100
Emergency Net of Martin Co.	fm	147.06	F	0100
Fast Net	lsb	3.940	Dy	0330
Fast Net	lsb	3.940	Dy	2300
Florida Crown Emergency Net, northeast Florida	fm	146.76	Th	0030
Florida Midday Traffic Net	lsb	7.247	Dy	1700
Florida Phone Traffic Net	lsb	3.940	Dy	1200
Jumpsuit Net	fm	14.180	M-F	2000
Lee Co. ARES Net	fm	146.88	F	0030
Lee Co. RACES Net	fm	146.88	2nd4th Th	0100
Lemon Bay Emergency Net, Englewood	fm	146.775	W	2330
Manatee Co. ARES/RACES Net	fm	146.925	TF	0100
Marion County Emergency Net	fm	146.61	T	0030
Marion County RACES Net	fm	146.61	Th	0030
Milwaukee-Florida Net	usb	14.290	Dy	1400
Northern Florida Phone Net	lsb	3.950	Dy	2330
Orange County ARES/RACES	fm	146.82	F	0000
Osceola Co. ARES/RACES Net	fm	146.79	T	0030
Palm Beach ARES Traffic Net	fm	146.67	M-Sa	0000
Pensacola Area Emergency Net	fm	146.76	T	0000
Pinellas Co. ARES/RACES Emergency Net Florida	fm	147.06	Th	0030
Pinellas Co. ARES/RACES Emergency Net Florida	fm	147.36	Th	0030

Net	Mode	Freq	Day	Time
Polk County ARES Net	fm	146.685	W	0030
QFN All Florida CW Traffic Net	cw	3.651	Dy	0000
QFN All Florida CW Traffic Net	cw	7.061	Dy	0300
S. Brevard Technical, Emergency & Ragchew Net	usb	29.200	Dy	0030
Santa Rosa Co. Area Emergency Service Net	fm	146.70	M	0030
Sarasota Emergency Radio Club	fm	146.73	Th	0030
Spanish Dade Emergency Net, Miami	fm	147.15	F	0100
Tropical Phone Traffic Net	lsb	3.940	Dy	2200
Tropical Phone Traffic Net	lsb	7.247 alt.	Dy	2200
Volusia Emergency Net	fm	147.42	Dy	0000
West Pasco Co. ARES	fm	146.67	W	1900

Georgia

Net	Mode	Freq	Day	Time
135 Repeater Group	fm	147.135	Th	0200
ARC of Augusta Net	fm	146.985	M	0100
ARC of Savannah 2-Meter Net	fm	146.97	Th	0200
Augusta Net	fm	146.985	M	0100
Big Shanty-135 Repeater Group, Atlanta	fm	146.655	W	0200
Big Shanty-135 Repeater Group, Atlanta	fm	147.135	W	0200
Big Shanty-135 Repeater Group, Atlanta	fm	224.12	W	0200
Carroll Co. ARES	fm	146.64	T	0100
Chatham Co. Emergency ARC Net	fm	147.015	Th	0100
Chattahoochee Emergency Net, west central Georgia	fm	146.61	Th	0130
Chattanooga Area Weather Net	fm	146.61	W	0100
Clayton ARES Net	fm	145.17	M	0130
Coastal Area Repeater Society	fm	146.70	Sn	0200
Conyers Amateur Radio Group ARES, Covington	fm	147.21	M	0100
Coweta-Heard County ARES	fm	145.53	Sn	0200
Dalton Amteur Radio Club	fm	145.23	M	0230
Dalton ARC ARES	fm	145.23	S	0300
DeKalb Amateur Radio Emergency Service	fm	145.45	M	0200
Georgia ARES Administration	lsb	3.975	Sn	2200
Gwinnett Co. Emergency Net	fm	147.075	W	0100
Liberty Co. Emergency ARC Net	fm	147.015	Th	0100
Metro Atlanta Emergency Net	fm	145.47	M	0200
Metro Atlanta Emergency Net	fm	145.47	S	1500
North Georgia 220 Net	fm	224.74	on call	
North Metro Emergency Net, north Georgia	fm	145.47	S	1500
North Metro Emergency Net, north Georgia	fm	145.47	T	0000
Northwest Georgia ARES Net	fm	146.715	S	0300
Tri State Two Meter Net	fm	147.300	Sn,T,Th	2130
Wayne Co. Emergency ARC Net	fm	147.015	Th	0100

Hawaii

Net	Mode	Freq	Day	Time
Big Island Emergency Net	lsb	3.905	Th	0600
Big Island Emergency Net	usb	28.305	M	0500
Big Island Emergency Net	fm	146.76	Th	0500
Emergency ARC 2-Meter Net	fm	146.88	Dy	0530
Friendly Net, statewide	lsb	7.285	Dy	1900
Hawaii Afternoon Net	lsb	7.290	Dy	0200
Hawaii Emergency Net	lsb	7.290	on call	
Kauai ARC VHF Net	fm	146.91	T	0600
Maui Emergency Net	cw	7.120	T	0700
Maui Emergency Net	lsb	7.2545	1st T	0515
Maui Emergency Net	fm	147.02	T	0500

Idaho

Net	Mode	Freq	Day	Time
Eagle Rock RACES-ARES NET, east and southeast Idaho	fm	146.64	W	0400

Friendly Amateur Radio Missions Net	lsb	3.937	Dy	0200
Idaho Civil Defense Net	lsb	3.990	M-F	1500
Mini-Cassia CD Net, Burley-Rupert	fm	146.52	S	0300
Nevada Weather Net	lsb	3.993	M-Sa	0600
Northwest Traffic Net	fm	146.98	Dy	0200
Northwest Traffic Net	fm	147.22	Dy	0200
Treasure Valley Emerg Net, Boise/Nampa	fm	146.94	S	0130
Twin Falls County ARES	fm	146.66	W	0230
Voice of Idaho 2-Meter Emergency Net, southwest Idaho	fm	147.24	F	0300

Illinois

Adams Co. Emergency Service Net	fm	146.94	Th	0100
Adams Co. Emergency Service Net	fm	147.03	Th	0105
Argonne ARC Net, Metro Chicago	fm	145.19	T	0300
Boone Co. Emergency Net	fm	147.375	T	0145
Cass County ARES Net	fm	146.715	M	0300
Central Illinois Emergency Net, Sangamon Co.	fm	146.61	F	0200
Champaign County ARES Net	fm	146.52	3rd M	0300
Champaign County ARES Net	fm	147.06	1st M	0300
Chicago FM Club Info Net	fm	146.76	W	0200
ChicagoLink repeater system	fm	29.68	M	2330
ChicagoLink repeater system	fm	52.825	M	2330
ChicagoLink repeater system	fm	145.45	M	2330
ChicagoLink repeater system	fm	147.075	M	2330
ChicagoLink repeater system	fm	147.225	M	2330
ChicagoLink repeater system	fm	223.90	M	2330
ChicagoLink repeater system	fm	224.16	M	2330
ChicagoLink repeater system	fm	224.48	M	2330
ChicagoLink repeater system	fm	224.80	M	2330
ChicagoLink repeater system	fm	444.75	M	2330
Christian Co. ARES Net	fm	146.955	F	0000
DeKalb Area AR Net	fm	146.73	T	0300
DuPage Co. ARES	fm	146.61	T	2000
FRRL/NIARC 2-Meter Net, northeast Illinois	fm	146.58	W	0130
Gypsies Info Net, Will Co.	fm	146.82	F	0300
Heart of IL FM Repeater Disaster Net	fm	146.76	W	0230
Illinois ARES Net	lsb	3.905	1st,3rd Sn	2230
Illinois ARES Net, Carbondale	fm	146.64	1st,3rd Sn	2230
Illinois ARES Net, northeast Illinois	fm	147.06	1st,3rd Sn	2230
Illinois Emergency Net	lsb	3.940	S	1400
Kane Co. ARES Net	fm	145.47	on call	
Kendal Co. Net	fm	147.375	Th	0130
Knox Co. Emergency Traffic & Training Net	fm	147.21	M	0300
La Moine Emergency AR Net	fm	147.06	on call	
Lake Co. RACES	fm	147.18	T	0200
Lee Co. RACES Net	fm	146.97	TF	0200
Macon Co. ARES Net	fm	146.73	T	0200
Madison Co. ARES Net	fm	145.13	T	0100
McLean Co. ARES Net	fm	146.79	on call	
North Central Phone Net	lsb	3.913	M-Sa	13+1800
Northeast Illinois Emergency Net	fm	147.06	T	0200
Northwest ARES, Schaumburg	fm	147.285	1st S	0130
Northwest ARES, Schaumburg	fm	443.625	1st S	0130
Okaw Valley ARC Net, Greenville	fm	147.24	M	0300
Piatt Co. ARES	fm	146.925	T	0230
Radio Net, Wabash Co.	fm	146.94	W	0300

Schaumburg ARC	fm	144.600	Th	0200
Schaumburg ARC	usb	28.350	Th	0200
Six Meter Club of Chicago	fm	146.97	W	0300
South Jacksonville Repeater Network	fm	147.00	W	0200
Southern Illinois ARS	fm	146.64	W	0300
Streator ARES Net	fm	146.52	M	0200
Valley Amateur Repeater Net, Elgin	fm	146.79	T	0300
VARC ARES Net, Metro Chicago	fm	146.73	M	0200
W9AML-Tuesday Nite 2 Meter Net, Bloomington	fm	146.94	W	0130
W9AML-Tuesday Nite 2 Meter Net, Normal	fm	146.94	W	0130
W9AML-Tuesday Nite 2 Meter Net, Peoria	fm	146.94	W	0130
W9VEY Memorial 2-Meter Net, Hillsboro	fm	146.82	T	0230
Whiteside Co. ARES Net	fm	145.21	alt M	0200
Whiteside Co. ARES Net	fm	146.85	alt M	0200
Winnebago Co. ARES Emergency Net	usb	28.375	T	0245
Winnebago Co. ARES Net	fm	146.61	T	0230
Woodford Co. 2-Meter Volunteer Net	fm	147.27	W	0300

Indiana

21 Repeater Group, Whitley Co.	fm	147.15	Dy	0200
Boone Co. ARC Net	fm	147.105	F	0000
Cass Co. ARC Net	fm	147.18	T	0230
ChicagoLink repeater system, northeast Indiana	fm	29.68	M	2330
ChicagoLink repeater system, northeast Indiana	fm	52.825	M	2330
ChicagoLink repeater system, northeast Indiana	fm	145.45	M	2330
ChicagoLink repeater system, northeast Indiana	fm	147.075	M	2330
ChicagoLink repeater system, northeast Indiana	fm	147.225	M	2330
ChicagoLink repeater system, northeast Indiana	fm	223.90	M	2330
ChicagoLink repeater system, northeast Indiana	fm	224.16	M	2330
ChicagoLink repeater system, northeast Indiana	fm	224.48	M	2330
ChicagoLink repeater system, northeast Indiana	fm	224.80	M	2330
ChicagoLink repeater system, northeast Indiana	fm	444.75	M	2330
Clark Co.	fm	146.85	Th	0200
Clifty Emergency Net, Jefferson Co.	fm	146.52	F	0100
Clinton Co. ARES Net	fm	146.61	T	0100
Delaware Co. RACES Net	fm	146.73	T	0200
Du Boise Co. ARES Net	fm	147.195	F	0130
East Central Indiana Repeater Net	fm	147.00	T	0200
Elkhart Co. ARES/RACES Net	fm	145.43	M	0030
Fayette Co. ARES Net	fm	146.745	W	0200
Floyd Co. RACES Net	fm	146.85	1st,3rd W	0300
Gibson Co. ARES Net	fm	145.41	W	0200
Grant Co. ARES Net	fm	146.79	Dy	0020
Hamilton County ARES/RACES Net	fm	147.39	M	0100
Hancock Co. ARES Net	fm	145.30	W	0030
Hendricks Co. ARES/RACES	fm	147.03	W	0200
Hendricks Co. ARES/RACES	usb	28.400	T	0200
Henry Co. 2 Meter RACES Net	fm	145.27	Th	0100
Henry Co. 10-Meter RACES Net	usb	28.340	T	0130
Huntington Co. ARES Net	fm	146.685	W	0100
Indiana ARES Net	lsb	3.910	last Sn	2315
Indiana RACES Net	lsb	3.9965	3rd S	1500
Indiana Traffic Net	lsb	3.910	Dy	1330
Indiana Traffic Net	lsb	3.910	Dy	2130
Indiana Traffic Net	lsb	3.910	Dy	2300
Indiana Wet Net	lsb	3.910	Dy	1310

Indiana, Michigan and Ohio Net, Allen Co.	fm	146.88	Dy	2330
Indianapolis Repeater	fm	146.70	Dy	24 hours
Jackson Co. ARES Net	fm	145.43	M	0030
Johnson Co. ARES Net	fm	146.835	M	0100
Kokomo 2FM Wet Net	fm	146.91	Dy	1255
Kokomo ARES Net	fm	146.91	M-F	1600
Kosciusko ARES Net	fm	146.385	W	2330
Lake Co. ARES Net	fm	147.00	T,Th,S	0030
LaPorte Co. Emergency Net	fm	146.61	S	1445
Marion Co. ARES Net	fm	146.70	Th	0030
Marion Co. RACES Net	fm	146.76	W	0000
Marshall Co. SKYWARN Net	fm	147.285	on call	
Martin Co. ARES Net	fm	145.21	M	0130
Miami Co ARES/RACES Net	fm	147.345	M	0130
Mid-State ARC Net, central Indiana	fm	146.835	M	0000
Monroe Co. ARES Net	fm	146.64	T	0030
Morgan Co. ARES Net	fm	147.06	W	0230
North Central Indiana Traffic Net, Howard Co.	fm	147.91	M-F	0045
North Central Phone Net	lsb	3.913	M-Sa	13+1800
Northeast Illinois Emergency Net	fm	147.06	T	0200
Old Post ARES Net, Knox Co.	fm	146.67	F	0100
Parke Co. EC Net	fm	147.30	T	1900
Porter Co. ARES Net	fm	146.775	Th	0200
Putnam Co. ARES Net	fm	147.33	Sn	1400
Randolph Co. RACES NET	fm	147.30	3rd S	1500
Ripley Co. ARES Net	fm	146.805	Sn	0100
Ripley Co. Repeater Net	fm	146.805	W	0100
Southeast Indiana RACES Net, Clarke Co.	fm	146.85	1st,3rd W	2130
St. Joseph Co. ARES Net	fm	147.39	T	0000
St. Joseph Co. SKYWARN Net	fm	147.39	on call	
Sullivan Co. Emergency Net	fm	146.52	F	0000
Tioga ARS Net, White Co.	fm	147.075	M	0000
Tippecanoe Co. ARC Net	fm	147.135	F	0030
Tri-Co ARES Net, Park Co.	fm	147.30	T	1900
Tri-State Emergency Net, Vanderburg Co.	fm	147.15	Th	0200
Vigo Co. Emergency Net	fm	147.30	W	0030
Wabash Valley ARES Net, Vigo Co.	fm	146.85	Sn	1330
Warrick Co. Net	fm	147.075	M	0200
Washington County ARES	fm	146.655	T	0030
Wayne Co. ARES Net	fm	147.27	Th	0100
Whitley Co. ARES Net	fm	147.15	Th	0015

Iowa

Cass Co. ARES Net	fm	147.15	T-S	0100
Central Iowa District 1 ARES Net, Polk Co.	fm	146.67	Sn	0000
Clinton Co. ARES Net	fm	145.43	M	0230
Fort Dodge ARES Net, Webster Co.	fm	146.91	M-F	0030
Fort Madison ARES	fm	147.30	W	0200
Great Lakes ARES Net	fm	146.61	W	0100
Humboldt Area 2-Meter Net	fm	147.18	M-Th	0100
Iowa Traffic & Emergency Net	lsb	3.970	Sn	2330
Jasper Co. ARES	fm	147.03	Dy	0100
Johnson Co. Emergency Net	fm	146.85	M	0130
North Central Phone Net	lsb	3.913	M-Sa	13+1800
Northern Iowa ARES Net, north central Iowa	fm	146.76	T	0030
Northern Iowa ARES Net, north central Iowa	fm	147.15	M	0300

O'Brien Co. ARES Net	fm	146.52	W	0030
Osceola Co. ARES Net	fm	147.30	W	0110
Page Co. ARES Net	fm	146.97	Sn	0200
Polk Co. ARES Net	fm	146.94	F	0100
Pottawattamie Co. ARES Net	fm	146.82	Th	0300
Scott Co. ARES Net	fm	146.88	M	0200
Story Co. ARES	fm	146.76	S	0100
Wapsipinicon ARES Net, Buchanon Co.	fm	145.33	M	0000
Warren Co. ARES Net	fm	146.52	F	0100
West Central ARES Net	fm	147.09	M-F	0230

Kansas

Central Kansas ARC ARES Net	fm	147.03	Th	0200
Cowley Co 2 Meter Emergency Net	fm	146.70	M	0300
Johnson Co. Emergency Communications Service	fm	145.47	M	0030
Kansas Phone Net	lsb	3.920	M,W,F	1245
Kansas Phone Net	lsb	3.920	S,Sn	1400
Kansas Sideband Net	lsb	3.920	Dy	0030
Kansas Tri-Zone ARES Net, north central Kansas	fm	146.73	M	0300
Kansas Weather AM Net	lsb	3.920	Dy	1300
Kansas Weather PM Net	lsb	3.920	Dy	0000
Kansas Zone 1 ARES JARS Net, Wyandotte Co.	fm	147.15	Th	0100
Kansas Zone 2 ARES Net, Leavenworth Co.	fm	147.00	Th	0130
Kansas Zone 3 ARES Net, Johnson Co.	fm	146.46	S	0130
Kansas Zone 4 ARES Net, Miami Co.	fm	147.36	Su	0100
Kansas Zone 9 ARES Net, Greenwood Co.	fm	146.985	M	0300
Kansas Zone 9 ARES Net, Lyon Co.	fm	146.985	M	0300
Kansas Zone 12A ARES Net, Douglas Co.	fm	146.76	W	0300
Kansas Zone 12B ARES Net, Franklin Co.	fm	146.805	Th	0300
Kansas Zone 12B ARES Net, Osage Co.	fm	146.805	Th	0300
Kansas Zone 13 ARES Net, Jack. Co.	fm	146.67	W	0230
Kansas Zone 13 ARES Net, Pott. Co.	fm	146.67	W	0230
Kansas Zone 13 ARES Net, Shaw. Co.	fm	146.67	W	0230
Kansas Zone 13 ARES Net, Wab. Co.	fm	146.67	W	0230
Kansas Zone 18A ARES Net, Clay Co.	fm	147.06	M	0300
Kansas Zone 18C ARES Net, Geary Co.	fm	146.88	F	0400
Kansas Zone 19A ARES Net, Elisw	fm	147.03	Th	0200
Kansas Zone 19A ARES Net, Lincoln	fm	147.03	Th	0200
Kansas Zone 19A ARES Net, Lincoln	usb	28.450	T	0200
Kansas Zone 19A ARES Net, Ott	fm	147.03	Th	0200
Kansas Zone 19A ARES Net, Ott	usb	28.450	T	0200
Kansas Zone 19A ARES Net, Salinas	fm	147.03	Th	0200
Kansas Zone 19A ARES Net, Salinas	usb	28.450	T	0200
Kansas Zone 19A ARES Net, Wllsw	usb	28.450	T	0200
Kansas Zone 20A ARES Net, Dickinson Co.	fm	145.33	F	0200
Kansas Zone 27 ARES Net, Barton Co.	fm	146.76	Th	0300
Kansas Zone 32 ARES Net, Finney Co.	fm	146.91	Su	1915
Kansas Zone 32 ARES Net, Gray Co.	fm	146.91	Su	1915
Kansas Zone 32 ARES Net, Haskell Co.	fm	146.91	Su	1915
Kansas Zone 36 ARES Net, Osb.	fm	147.375	W	0300
Kansas Zone 36 ARES Net, Phillips	fm	147.375	W	0300
Kansas Zone 36 ARES Net, Rooks	fm	147.375	W	0300
Kansas Zone 36 ARES Net, Smith	fm	147.375	W	0300
Kansas Zone 37 ARES Net, Ellis Co.	fm	147.18	Tu	0100
Kansas Zone 37 ARES Net, Russel Co.	fm	147.18	Tu	0100
Kansas Zone 37 ARES Net, Trego Co.	fm	147.18	Tu	0100

Kansas Zone 39 ARES Net, Grove	fm	146.82	Th	0300
Kansas Zone 39 ARES Net, Logan	fm	146.82	Th	0300
Kansas Zone 39 ARES Net, Rawlins	fm	146.82	Th	0300
Kansas Zone 39 ARES Net, Thomas	fm	146.82	Th	0300
Kansas Zone 40 ARES Net, Cheyenne Co.	fm	146.82	Th	0300
Kansas Zone 40 ARES Net, Sherman Co.	fm	146.82	Th	0300
Kansas Zone 40 ARES Net, Wallace Co.	fm	146.82	Th	0300

Kentucky

Bullitt Co. ARES	fm	146.70	T	0100
Carter Co. Emergency Net	fm	146.70	W	0200
District 3 ARES Net	fm	147.21	F	0300
District 4 ARES Net	fm	146.85	F	0100
District 5 ARES Net, central Kentucky	fm	146.98	W	0100
District 6 ARES Net	fm	146.70	Th	0200
District 7 ARES Net, north Kentucky	fm	147.375	F	0000
District 11 ARES Net, southeast Kentucky	fm	146.61	M,Th	0130
District 14 ARES Net	fm	145.49	F	0130
Kentucky CW Net	cw	3.600	Dy	0100
Kentucky Traffic Net	lsb	3.959	Dy	0000
Letcher Co. Emergency Net	fm	147.39	Th	0100
Northern Kentucky Emergency Net	fm	147.255	W	0030
PAEWTN, west Kentucky	fm	147.06	M	0200
Pennyrile KY Area Weather Net	fm	146.61	T	0300
Southeast Kentucky Emergency Net-S, Hazard	fm	146.67	M	0200
Trimble Co. ARES	fm	146.64	on call	
Triple States ARES High-Speed CW Net	cw	28.480	Th	0010
Triple States ARES Intermediate CW Net	cw	28.480	W	2350
Triple States ARES Phone Net	usb	28.480	Th	0030
Tug Valley ARES/RACES Net	fm	145.330	M	0200
Wilderness Trail Emerg. Net, south central Kentucky	fm	146.717	Th	0230
Woodford ARES, central Kentucky	fm	145.575	F	0100

Louisiana

AR-LA-TX 10 Meter Net, northwest Louisiana	usb	28.400	Th	0330
Louisiana Traffic Net	lsb	3.910	Dy	0030
United Radio Amateur Club Net, northwest Louisiana	fm	146.94	Sn	0230
W5YL Thibodeaux ARC, LaRouche	fm	147.300	2nd4th5th T	0100
West Gulf Emergency Net	lsb	3.9435	Sn	1400

Maine

Augusta Area ARES Net	fm	146.85	M	0100
Bath-Brunswick Area ARES Net	fm	147.27	T	0030
Central Maine ARES Net	fm	146.70	W,S	0100
Maine Public Service Net	lsb	3.940	Sn	1400
Pine Tree Net	cw	3.596	Dy	0000
Presque Isle Area ARES Net	fm	146.73	T	0100
Sea Gull Net	lsb	3.940	M-Sa	2200
Southern Maine District ARES Net	fm	147.09	T	0100
York Area ARES Net	fm	146.775	T	0030

Maryland

Baltimore City/County ARES Net	fm	145.13	1st W	0030
BRATS Weather Net, Baltimore	fm	147.03	on call	
BRATS-Emergency Net, Baltimore	fm	147.03	Th	0100
Carroll Co. Amateur Radio Emergency Team	fm	145.41	T	0030
Carroll Co. Amateur Radio Emergency Team	usb	28.190	F	0030
Howard Co. ARES Net	fm	147.135	1st,3rd T	0030
Kent Co. ARES	fm	147.26	T	0100

Net	Mode	Frequency	Day	Time
Maryland Emergency Phone Net	lsb	3.920	Dy	2300
Maryland Packet Traffic Net	pkt	@ N2GTE-6	Dy	24 hours
Maryland-Delaware-DC Traffic Net	cw	3.643	Dy	0000
Maryland-Delaware-DC Traffic Net	cw	3.643	Dy	0300
MDC EC & RACES Officer Net	fm	147.135	M	0130
National Capital ARES, Washington DC	fm	146.91	M	0200
Prince George's Co. ARES	fm	147.15	W	0200
West Virginia Weather & Travelers Net	fm	145.270	T	0000
West Virginia Weather & Travelers Net	fm	147.285	T	0000
West Virginia Weather & Travelers Net	fm	444.400	T	0000
Wicomico Co. ARES	fm	146.82	T	0030
WVA, MD, VA Net	lsb	3.905	M-Sa	2200

Massachusetts

Net	Mode	Frequency	Day	Time
Aquidneck Island Communications Net, east Mass.	fm	147.36	Sn	1315
Cambridge ARES Net	fm	145.545	3rd T	0030
Cape & Islands Weather Net	fm	147.045	M-F	1200
Danvers Emergency Net		29.560	T	0000
Danvers Emergency Net	fm	146.94	T	0015
Foxboro Advisory Net	fm	147.375	F	0030
Framingham Area AR Emergency Net	fm	147.15	M	0030
Lincoln Emergency Net	fm	145.605	1st,2nd M	0100
Massachusetts CD Agency Net 1A, Boston	fm	146.64	1st M	0100
Massachusetts CD Agency Net 1B, Waltham	fm	146.64	1st M	0100
Massachusetts CD Agency Net 1C, Lincoln	fm	145.545	1st M	0000
Massachusetts CD Agency Net 1D, Tewksbury	fm	145.68	1st M	0100
Massachusetts CD Agency Net 1E, Lawrence		145.09	1st M	0100
Massachusetts CD Agency Net 1F, Beverly		145.09	1st M	0100
Massachusetts CD Agency Net 2A, Taunton	fm	147.135	1st M	0000
Massachusetts CD Agency Net 2B, Bridgewater	fm	147.18	1st M	0030
Massachusetts CD Agency Net 2C, Barnstable	fm	147.045	1st M	0000
Massachusetts CD Agency Net 2D, Sharon	fm	146.865	1st M	0030
Montachuset Emerg. Net, north central Massachusetts	fm	145.45	Sn	1400
Mt. Tom Emergency Net, central Massachusetts	fm	146.94	Sn	1400
New England Emergency Phone Net	lsb	3.945	Sn	1330
Provin Mt. Emerg. Net, south central Massachusetts	fm	146.70	Sn	1400
Radio Amateurs Interstate Weather Net	fm	145.27	Dy	2300
So. Worcester ARES Net	fm	147.345	Sn	1400
So. Worcester Co. Emergency 2-Meter Net	fm	146.97	Sn	1400
So. Worcester Co. Emergency 220 Net	fm	223.74	Sn	1400 local
The RI ARES/Red Cross HF Net	usb	28.400	Sn	0000
Tri-State Emergency Net	fm	146.805	Th	0030
Twin State Emergency Net	fm	146.76	W	0030
Western Massachusetts Net	cw	3.562	Dy	0000
Western Massachusetts Emergency Net	lsb	3.937	M,Sn	1330
Western Massachusetts Emergency Net	lsb	3.937	M,Sn	2300
Western Massachusetts Emergency 6-Meter Net	fm	53.23	Sn	1415
Western Massachusetts Emergency 2-Meter Net	fm	146.91	Sn	1400
Western Massachusetts Emergency 220 Net	fm	224.10	Sn	1430
Winthrop Emergency Net	fm	447.40	1st T	0000
Worcester Emergency Net, central Massachusetts	fm	146.97	Sn	1400

Michigan

Net	Mode	Frequency	Day	Time
Berrien Co. ARES Net	fm	145.47	M	0000
Berrien Co. SKYWARN Net	fm	146.82	W	2330
Branch Co. ARES Net	fm	147.30	F	0010
Calhoun Co. Area 10 Meter Net	usb	28.350	T	0200

Calhoun Co. ARES 2 Meter Net	fm	147.12	Th	0200
Cass Co. ARES Net	fm	146.52	Th	0100
Chelsea Area Net	fm	146.98	T	0000
Chelsea SKYWARN	fm	146.90	on call	
Emergency Coordinator Organizational Net, s.w. Mich.	fm	147.12	Th	0030
Genesee Co. ARES Net	fm	147.26	F	1800
Genesee Co. ARES Net	fm	147.26	Th,F	0200
Great Lakes Emergency & Traffic Net	lsb	3.932	Dy	0200
Holland ARC Net	fm	147.06	M	2330
Ionia Co. ARES/RACES and Information Net	fm	145.13	T	0000
Jackson Action 2 Meter Net	fm	146.88	F	0200
Jackson Co. ARES/RACES Training Net	fm	146.88	M	0200
Kalamazoo Co. ARES/RACES Net	fm	147.04	Th	0100
Leelanau Co. ARES Net	fm	146.85	T	0040
Livingston Amateur Radio Klub Net	fm	146.68	M	0200
Macomb Co. RACES/ARES	fm	145.56	F	0100
Macomb Co. RACES/ARES	fm	147.20	T	0100
Manistee Co. Emergency Net	fm	146.78	M	0100
Michigan Emergency Net	lsb	3.930	Sn	1400
Michigan Net	cw	3.663	Dy	0300
Michigan Net	cw	3.663	Dy	2300
Michigan Net	cw	3.663	Dy	2330
Michigan Section ARPSC Net	lsb	3.932	Sn	2200
Michigan Section ARPSC Net	lsb	7.232 alt.	Sn	2200
Michigan Thumb Net, Sanilac Co.	fm	146.43	Th	0200
Michigan Traffic Net	lsb	3.953	Dy	0000
Monday Evening Amateur Social Hour, northwest Mich.	fm	146.85	T	0100
Monroe Co. ARES Net	fm	146.72	T	0200
Oakland Co. ARPSC/RACES Net	fm	147.16	Th	0100
Ottawa Co. ARES/RACES Net	fm	146.55	1st F	0030
Seventh District North Net, northwest Michigan	fm	146.82	2nd,4thTh	0100
Seventh District South Net, northwest Michigan	fm	146.97	2nd,4thTh	0200
Southeastern Michigan ICO	fm	224.94	on call	
State of MI District #3 Public Service Net	fm	145.31	Sn	2300
Sunday Morning Amateur Social Hour	lsb	3.935	Sn	1400
The Sunrise Net, west Michigan	fm	146.78	Dy	1100
Washtenaw Co. ARES Net	fm	146.58	T	0115
Washtenaw Co. RACES Net	fm	146.92	W	0100
Wayne Co. ARES Net	fm	145.33	Th	0200
Wayne Co. RACES Net	fm	147.14	Th	0145
West Michigan EC Roundtable, southwest Michigan	fm	147.165	1st Th	0200
Minnesota				
Arrowhead Ragchew Net, statewide	lsb	7.250	Dy	0030
Bemidji ARES Net	fm	146.73	M	0300
Carver/Scott Cos. ARES Net	fm	147.165	M	0230
Goodhue Co. Emergency Net	fm	147.30	M	0200
Mankato Emergency Services Net	fm	147.24	M	0330
Marshall Area Emergency Net	fm	147.195	Th	0300
Mille Lacs County ARES Net	fm	146.61	W	0200
Minnesota Amateur Weather Net	lsb	3.860	Dy	0015
Minnesota Section Net 1	cw	3.685	Dy	0030
Minnesota Section Phone Net	lsb	3.860	Dy	2330
Minnesota SKYWARN Emergency Weather Net	lsb	3.860	S	
New Ulm ARC ARES Net	fm	147.33	M	0300
Northern St. Louis Co. ARES, Mesabi Iron Range	fm	147.15	M	0200

Northland Weather Group	fm	147.18	M	0130
Northshore Repeater Association Net, Lake Superior	fm	147.09	M	0100
Otter Tail Co. ARES/SKYWARN Net	fm	146.64	Th	0300
Pico Net	lsb	3.925	on call	
Ramsey Co. Emergency Net	fm	147.12	Th	0300
Rochester ARES Net, Olmsted Co.	fm	146.82	M	0130
SKYWARN Emergency Net	fm	147.06	on call	
SKYWARN Emergency Net	fm	147.18	on call	
SKYWARN Emergency Net	lsb	3.860	on call	
Southwest Minnesota Emergency Services Net	fm	146.67	W	0045
St. Louis County Emergency Services Net	fm	146.94	M	0330
St. Peter RC Emergency Services Net	fm	147.135	M	0230
Willmar Area Emergency Services Net	fm	146.91	W	0100

Mississippi

Coast ARES Net, Gulf Coast	fm	146.73	Th	0100
Gulf Coast Emergency Net	fm	146.73	W	0100
Gulf Coast Sideband Net	lsb	3.925	Dy	2330
Hattiesburg Area Emergency Net	fm	146.52	M	0200
Hattiesburg Area Emergency Net	fm	147.36	M	0200
Lauderdale Co. ARES Net	fm	146.70	W	0100
Miss-Lou Emergency Net, west Mississippi	fm	147.27	M	0300
Neshoba ARC Emergency Net	fm	147.33	Sn	0200
Northeast Mississippi Weather Net	fm	147.38	M	0100
Pine Belt Emergency Net, central Mississippi	fm	146.61	M	0000
Rankin Co. ARES	fm	147.15	F	0200

Missouri

Amateur Radio Emergency Service Net, Audrain Co.	fm	147.255	F	0300
ARES Net, Jackson Co.	fm	146.97	S	1500
Central Missouri Emergency Net	fm	146.76	Th	0300
Christian Co. Emergency Net	fm	147.36	T	1830
Clay Co. ARES Net	fm	146.79	Th	0200
Hickory Co. Emergency Net	fm	147.255	Sn	0030
Hickory Co. RACES/ARES	lsb	3.963	M	0000
Indian Foothills ARC, northwest and central Missouri	fm	147.165	Th	0130
Macon County ARES	fm	146.805	F	0230
Mercury Amateur Radio Association	lsb	3.994	S	0230
Missouri Emergency Net	lsb	3.963	on call	
Missouri Emergency Net	lsb	7.263 alt.	on call	
MO Emergency Operations and Weather Net	lsb	3.963	Dy	2330
Northeast Missouri Emergency Net	fm	145.13	M,W,F	0130
Old Hickory 2-Meter Emergency Net, Hickory Co.	fm	146.52	Sn	1930
Pike & Lincoln Counties ARES Net	fm	145.19	Th	0130
Platte Co. ARES Net	fm	147.33	W	0125
Sedalia Amateur Repeater Net	fm	147.03	W	0300
Southwest Missouri ARES Net, Springfield	fm	146.91	T	0100
Southwest Missouri SKYWARN Net	fm	146.91	W	0100
St. Charles ARC Emergency Net	fm	146.67	T	0130
St. Charles County ARES Net	fm	145.49	T	0230
St. Louis ARES Net	fm	146.91	T	0200
Stone-Barry-Taney Co's ARES, Tablerock Lake	fm	147.345	M	0300
WAØVXG ARES Associated Net, northwest Missouri	fm	146.85	S	1430
Zero-Beaters ARES Emergency Net, east central Missouri	fm	147.25	W	0200

Montana

Big Sky Net, statewide	lsb	7.240	M,T,Th	1600
Capitol City Radio Club Emergency Net, w.central Mont.	fm	147.22	T	0230

Hellgate ARC Net, northwest Montana	fm	147.04	W	0300
Missoula Area Emergency Net, northwest Montana	lsb	3.910	Sn	1600
Montana RACES	lsb	3.947	1st,3rd Sn	1600
Montana Traffic Net	lsb	3.900	Dy	0030
Valley Co ARES Net	fm	147.37	W	0200

Nebraska

Blue Valley ARES Net, Polk Co.	fm	147.27	M,Th	0300
Blue Valley ARES Net, Seward Co.	fm	147.27	M,Th	0300
Blue Valley ARES Net, York Co.	fm	147.27	M,Th	0300
Buzzards Roost ARES Net, northeast Nebraska	fm	146.79	Th	0100
Eastern Nebraska 2-Meter ARES Net, Lincoln	fm	146.76	M-F	0300
Hamilton County ARES	fm	147.18	T	0200
Mid-Nebraska ARES Net, Grand Island	fm	146.94	Dy	0100
Midlands ARES Net, Omaha	fm	146.94	M	0200
Nebraska 40 Meter Net	lsb	7.282	Dy	1900
Nebraska Cornhusker Net, statewide	lsb	3.980	Dy	1830
Nebraska Morning Phone Net	lsb	3.983	Dy	1330
Nebraska Storm Net	lsb	3.982	Dy	0030
Pawnee ARC 2-Meter Net, Columbus	fm	146.64	T-S	0230
Scotts Bluff ARES 6 Meter Net		50.150	M	0230
Scotts Bluff ARES Net	fm	145.47	F	0300

Nevada

Clark Co RACES Net	fm	147.18	M	0115
Douglas Co ARACES Net	fm	147.33	M	0230
Mercury Amateur Radio Association	lsb	3.983	S	1500
Nevada State RACES Headquarters	lsb	7.2485	Th	1800
Nevada State RACES Net	lsb	3.9965	M, Th	0300
Nevada Weather Net	lsb	3.993	M-Sa	0600
Washoe County RACES Net	fm	147.18	T	0230
Western Nevada ARES Net	fm	146.61	F	0300

New Hampshire

Connecticut Valley FM Assn. Weather Net	fm	146.76	M-F	1200
Connecticut Valley FM Assn. Weather Net	fm	146.76	M-F	2200
Connecticut Valley FM Assn. Weather Net	fm	146.76	T-S	0300
Manchester ARES Net	fm	147.255	Sn	0130
Seacoast Emergency Net	fm	147.57	1st Th	0200
Seacoast Emergency Net	fm	146.805	other Th	0200
Seacoast Net, Portsmouth	lsb	3.895	Sn	1500
Tri-State Emergency Net	fm	146.805	Th	0030
Tri-State FM Emergency Net, Cheshire Co.	fm	147.205	W	0030
Twin State Emergency Net	fm	146.76	W	0030
Western Rockingham County Emergency Net	fm	146.85	Th	0030

New Jersey

Albany Emergency Service Net	fm	147.12	W	0030
Amateur Radio Highway Net, north New Jersey	fm	146.895	M-F	1155
Amateur Radio Highway Net, north New Jersey	fm	146.895	M-F	2120
Bulington Co. ARES	usb	28.450	W	2345
Burlington Co. ARES	cw	28.150	Th	0015
Burlington Co. Emergency Net	fm	145.47	Th	0000
Burlington Co. Emergency Net	fm	147.15	Th	0000
Camden Co. ARES	usb	28.400	T	0030
Camden County ARES	fm	146.82	T	0000
Englewood ARES Net	fm	147.135	1st,3rd Th	0100
Gloucester Co. ARES Net	fm	147.18	M	0100
Gloucester Co. ARES Net	fm	224.66	M	0100

Net	Mode	Freq	Day	Time
Hudson Co. Area Traffic & Emergency Net	fm	145.43	F	0030
Hudson Co. Area Traffic & Emergency Net	fm	224.28	F	0030
Jumpsuit Net	usb	14.180	M-F	2000
Mercer Co. Emergency Net	fm	146.67	4th F	0030
Moorestown Emergency Net	fm	145.555	W	0200
Morris Co. ARES Net	fm	146.895	2nd,3rd,4th T	0100
Morris Co. RACES Net	fm	146.895	1st T	0100
New Jersey Net	cw	3.695	Dy	0000
New Jersey Net	cw	3.695	Dy	0300
New Jersey Phone Net	lsb	3.950	Dy	2300
Ramapo Valley Emergency Net	fm	145.62	on call	
Ramapo Valley Emergency Net	fm	146.70	2nd, 4th W	0100
Severe Weather Amateur Radio Net	fm	146.445	on call	
Somerville ARES/RACES Net	fm	145.65	M	0100
South Jersey ARES Net	fm	147.345	T	0100
Southern New Jersey ARES Staff Net	fm	147.345	1,2,3 T	0100

New Mexico

Net	Mode	Freq	Day	Time
Bernlillo Co. ARES Net	cw	3.639	Th	1900
Bernlillo Co. ARES Net	fm	146.94	Th	1900
Curry Co. ARES Net	fm	147.24	M	0300
Dona Ana Co. ARES Net	fm	146.64	Sn	0230
Mercury Amateur Radio Association	lsb	3.994	S	0230
Mimbres ARES Net, Deming	fm	146.82	M	0130
New Mexico ARES/RACES Net	lsb	3.939	1st F	0145
Otero Co ARES Net	fm	146.80	S	0200
San Miguel Co. ARES Net	fm	147.30	F	0230
Sandoval Co. ARES Net	fm	147.10	T	0200
Socorro Co. ARES Net	fm	146.68	T	0300

New York

Net	Mode	Freq	Day	Time
Albany Co. ARES/RACES	fm	147.12	Th	0030
Albany Traffic & Emergency Net	cw	3.737	T,Th	0030
Babylon ARES Daytime Net	fm	146.685	M	1400
Babylon ARES Net	fm	146.685	T	0100
Big Apple VHF Traffic Net	fm	145.350	Dy	0100
Central District ARES	fm	147.00	3rd M	0130
Central New York Traffic Net	fm	147.00	Dy	0215
Central New York Traffic Net	fm	147.30	Dy	0215
Champlain Valley ARC Simulated Emergency Net	fm	147.15	Sn	1430
Chemung Co. ARES Net	fm	147.36	M	0200
Chenango Co ARES Net	fm	146.685	Th	0030
Clinton Co Emergency Traffic Net	fm	147.15	M-F	0000
Clinton, Essex, Franklin ARES/RACES Net	lsb	3.955	S	1400
Columbia Co. ARES	fm	147.21	W	0000
Cortland Co. ARES	fm	147.825	T	0100
County of Orange Volunteer Emerg. Service Training Net	fm	146.76	M,Th	0100
Essex County RACES	lsb	3.955	S	1500
Glens Falls ARES	fm	146.73	Sn	
Greene Co. ARES	fm	147.57	F	0000
Herkimer Co. ARES Net	fm	146.82	Th	0230
Jumpsuit Net	ssb	14.180	M-F	2000
Larkfield Emergency Preparedness Net, Long Island	fm	147.21	Th	0030
Larkfield Mariners Net, Long Island	fm	147.21	T	0100
Lewis Co. RACES/ARES Net	fm	147.015	Sn	2300
MAARC Simulated Emergency Net, New York City	fm	147.36	W	0130
Madison Co. ARES Net	fm	147.105	M	0100

Net	Mode	Freq	Day	Time
Mariners Net, Larkfield, Long Island	fm	146.805	Sn-M	0030 local
Mariners Net, Larkfield, Long Island	fm	147.21	W	0100
New York Phone Net	lsb	3.925	Dy	1800
New York Phone Net	lsb	7.230	Dy	1800
New York Public Operations Net	lsb	3.913	Dy	2200
New York Public Operations Net	lsb	3.925	Dy	2200
New York State CW Net	cw	3.677	Dy	0000
New York State CW Net	cw	3.677	Dy	0300
New York State CW Net	cw	3.677	Dy	1500
New York State CW Net	cw	7.040	Dy	0300
New York State CW Net	cw	7.040	Dy	1500
New York State Phone Traffic & Emergency Net	lsb	3.925	Dy	0230
New York State RACES Net	cw	3.530	Sn	1430
New York State RACES Net	cw	7.1025	Sn	1430
New York State RACES Net	lsb	3.9935	Sn	1400
New York State RACES Net	lsb	7.245	Sn	1400
Niagara Co. ARES	fm	146.955	F	0000
NYC ARES Net	fm	145.35	Dy	0100
Oneida Co. RACES/ARES Net	fm	146.94	1st,3rd W	0030
Oneida Co. Traffic & Emergency Net	fm	146.88	Dy	0230
Oneida Co. Traffic & Emergency Net	fm	146.94	Dy	0230
Oneida Co. Traffic & Emergency Net	fm	146.94	Dy	2330
Oneonta ARC Net, Otsego Co.	fm	146.85	Th	0100
Onondaga Co. Radio Emergency Net	fm	147.30	1st,3rd T	0030
Ontario Co. ARES/RACES Net	fm	146.82	Sn	1330
Orange Co. ARES	fm	146.76	1st Th	0100
Oswego Co. RACES/ARES Net	fm	147.15	M	0000
Putnam Co. ARES	fm	145.135	2nd W	0100
Rensselaer Co. ARES	fm	146.76	Th	0030
Rensselaer Co. ARES/RACES	fm	147.18	W	0030
Saratoga Co. ARES/RACES	fm	147.00	Sn	0200
Schenectady Co. ARES	fm	147.06	Sn	1830
Schenectady Co. ARES	lsb	3.953	Sn	1900
SKYWARN Net, central and western New York	fm	147.00	Sn	2345
SKYWARN Net, central and western New York	fm	147.015	Sn	2300
SKYWARN Net, Chautauqua Co.		144.29	on call	
SKYWARN Operations, Southern ARES District	fm	147.075	on call	
Southern Tier Amateur Radio Net	fm	146.73	Dy	2330
St. Lawrence Co. 220 Net	fm	224.74	F	0000
St. Lawrence County ARES Net	fm	146.91	M	0030
Steuben Co. ARES/RACES Net	fm	145.19	T	0100
Suffolk County ARES Net	fm	145.33	T	0200
Tioga ARES/RACES	fm	146.76	M	0100
Tompkins Co. Amateur Radio Club Swap Net	fm	146.97	M	0100
Warren Co. ARES	fm	146.73	T	0001
Warren/Washington Co. ARES/RACES	fm	146.73	Sn	0001
Warren/Washington Co. ARES/RACES	lsb	3.968	Sn	0130
Wayne Co. ARES/RACES Net	fm	146.685	Sn	1345
Westchester Amateur Radio Association Net	usb	28.820	W	0100
Western Catskills ARES Net, Delaware Co.	cw	3.540	M	2330
Western District Net	fm	146.64	Dy	0230
Western District Net	fm	146.64	Dy	1600
Western District Net	fm	146.64	Dy	2330
Western New York Emergency Net	fm	146.73	M	0000
Western New York Emergency Net	fm	146.91	F	0000

Western New York Emergency Net	lsb	3.915	Sn	1730
Western New York Section ARES Coordination	lsb	3.955	on call	

North Carolina

Buncombe Co. ARES Net, Asheville	usb	28.740	Sn	1400
Buncombe Co. RACES	fm	146.91	W	0200
Cape Fear ARS Net, Cumberland Co.	fm	146.91	Dy	0000
Charlotte SKYWARN Net	fm	145.17	on call	
Charlotte SKYWARN Net	fm	145.35	on call	
Coastal Carolina Emergency Net	lsb	3.907	Dy	0000
Davidson Co. ARES Net	fm	146.91	W	0200
Forsyth Co. ARES	fm	146.64	Th	0200
Greensboro SKYWARN Net	fm	146.79	W	0100
Jumpsuit Net	usb	14.180	M-F	2000
Metrolina Two-Meter Emerg. Net, Mecklenburg Co.	fm	145.29	Dy	0200
North Carolina ARES Net	lsb	7.232	on call	
Piedmont Coast Traffic Net, Raleigh	fm	146.88	Dy	0200
Piedmont Emergency Training Net	fm	145.35	T-M	0100
Pitt County Emergency Communications Net	fm	147.09	T	0200
Raleigh ARS 10-Meter Training Net	usb	28.364	T,W,F,S	0130
Raleigh SKYWARN Net	fm	146.88	on call	
Rockingham Co. Emergency Net	fm	147.39	Th	0215
Tar Heel Emergency Net	lsb	3.923	Dy	0030
Wilson Amateur Radio Association ARES Net	fm	146.76	T	0130

North Dakota

Dickinson ARES Net, southwest North Dakota	fm	146.82	W	0200
Forx ARC Net, Grand Forks	fm	146.94	T	0300
Forx Area Net, Grand Forks	fm	146.94	M	0200
Forx ARES Net, Grand Forks	fm	146.94	M	0230
Goose River Two Meter Net, east central North Dakota	fm	146.91	Dy	1430
MonDak Net, west North Dakota	fm	146.73	F	0200
North Dakota Traffic Net	lsb	3.937	D	0030
North Dakota Weather Net (summer)	lsb	3.937	M-F	1500
North Dakota Weather Net (winter)	lsb	3.937	M-F	1830
North Forty Net, west North Dakota	fm	146.64	M	0300
Red River Radio Amateurs Net, Fargo	fm	146.76	Sn	0300
SKYWARN Emerg. Net, Storm Net	lsb	3.937	on call	
SKYWARN Emergency Net, Fargo	fm	147.06	on call	
Super Link, statewide	fm	145.0-147.0	M	0100

Ohio

ALERT, north central Ohio	fm	147.30	W	2000
Athens County Emergency Communications Net	fm	146.94	M	2000 local
Buckeye Net	cw	3.577	Dy	0300
Central Ohio ARES 2-Meter Net	fm	147.06	W	0100
Champaign-Logan ARC Emergency Net	fm	147.00	T	2030 local
Clyde Amateur Radio Society, northwest Ohio	fm	145.35	M	2100
Columbiana County Emergency Training Net	fm	146.805	M	2000 local
Cuyahoga Co. ARES Net	fm	145.29	M	2100
DARA Van Communications Net, Dayton	fm	146.94	M	2100
Grandfathers Net	lsb	3.9625	M-F	1330
Green County Traffic & Info Net	fm	147.165	W	2030
Hamilton Co. ARPSC	fm	146.46	1st F	2100
Hancock Emergency AR Services	fm	147.15	Th	2000 local
LEARA SKYWARN Net, northeast Ohio	fm	146.76	on call	
Licking Co. ARES	fm	146.88	W	1945
Lucas Co. Ohio ARES Net	fm	146.94	Dy	1845

Net	Mode	Freq	Day	Time
Mansfield Amateur Service & Emergency Repeater Net	fm	146.94	M	2000 local
Mercer Co., Pa., ARES/RACES	fm	147.15	F	0200
Miami County Emergency Services	fm	145.23	Sn	2100 local
Miami Valley Emergency Net		1.820	Sn	0900
Miami Valley FM Association WARN	fm	146.64	on call	
Northern Panhandle/Wheeling Emergency Net	fm	146.76	T	0115
Ohio Section ARES Net	lsb	3.875	Sn	2200
Ohio Section ARES Net	lsb	3.875	Sn	1700
Ohio Single Sideband Net	lsb	3.9725	Dy	1530
Ohio Single Sideband Net	lsb	3.9725	Dy	2115
Ohio Single Sideband Net	lsb	3.9725	Dy	2345
Ottawa Co. ARES	fm	147.075	T	2100
Portage Co. ARES Emergency Net	fm	145.68	W	2000 local
Portage Emergency Net	fm	145.39	Sn	2100 local
Queen City Emergency Net, Cincinnati	fm	147.24	T	2200 local
Rescue 40 Search & Disaster Team Net	fm	147.165	Th	0200
Stark Co. ARES Net	fm	147.12	W	1900
Steubenville-Weirton ARC Weather Net	fm	147.06	T	1945
Triple States Weather Net SKYWARN	fm	146.91	T	2015
Tuscarawas Co. ARC ARES Journal of the Air	fm	146.73	M	2000
Van Wert Area Emergency Net	lsb	3.920	Sn	1300
Warren County RACES	fm	146.865	1st,2nd,4th M	2030
Washington Co. ARES Net, Marietta	fm	146.88	Sn	2100
West Central Ohio ARES Net	fm	145.11	W	2000
West Virginia Weather & Travelers Net	fm	145.27	T	0000
West Virginia Weather & Travelers Net	fm	147.285	T	0000
West Virginia Weather & Travelers Net	fm	444.40	T	0000

Oklahoma

Net	Mode	Freq	Day	Time
Central Oklahoma District ARES Net	fm	146.82	Th	0200
Cimarron ARA Net, northwest Oklahoma	fm	145.45	F	0200
Edmond OK ARS Weather Net	fm	147.135	on call	
Eufaula Traffic & Emergency Net, east central Oklahoma	fm	145.37	Dy	2330
Grady/Caddo Co. ARES Net	fm	147.87	F	0100
Lawton Weather Training Net	fm	146.31	Th	0100
Mayes Co. ARES/RACES Net	fm	147.06	on call	
Mercury Amateur Radio Association	lsb	3.994	S	0230
Oklahoma Co. RACES HF Net, statewide	lsb	3.997	Sn	1300
Oklahoma Phone Emergency Net	lsb	3.900	Sn	1400
Oklahoma Traffic & Weather Net	lsb	3.900	M-Sa	2345
Pontotoc and Garvin Co. ARES Net	fm	147.285	T	0130
Salvation Army Net, central Oklahoma	fm	146.82	2nd S	1500
Tulsa ARES/RACES	fm	146.88	T	0300
Tulsa SKYWARN	fm	146.88	on call	

Oregon

Net	Mode	Freq	Day	Time
Central Oregon ARES	fm	147.06	Sn	0330
Clackamas ARES	fm	146.94	Sn	0300
Clatskanie Area ARES	fm	145.35	M	0300
Jackson Co. ARES	fm	146.94	Sn,T,F	0330
Lincoln Co. ARES	fm	145.37	W	0400
Lincoln-Benton-Linn Co. ARES	fm	146.78	Dy	0330
Marion-Polk Co. ARES	fm	146.86	M-Th	0300
Multnomah Co. ARES	fm	147.28	T	0300
Nevada Weather Net, Oregon	lsb	3.993	M-Sa	0600
Oregon ARES Traffic Net	lsb	3.9935	Dy	0115
Oregon Emergency Management Net	lsb	3.9935	Dy	1630

Oregon Emergency Net	lsb	3.980	Dy	0200
Oregon Section ARES Net	lsb	3.9935	on call	
Oregon Section EC Roundtable	lsb	3.9935	T	0400
Portland Area ARES	fm	147.32	Dy	0330
Tillamook Emergency Amateur Radio Service	fm	147.22	Th	0300
Umatilla/Morrow ARES	fm	146.80	2nd T	0300
Umatilla/Morrow ARES	fm	146.88	4th T	0300
Umatilla/Morrow ARES	fm	147.16	1st,3rd,5th T	0300
Washington Co. ARES	fm	147.04	W	0300
Western Lane ARES	fm	146.80	W	0300

Pennsylvania

Armstrong Co. ARES Net	fm	444.90	T	0200
Bedford Co. Emergency Preparedness Net	fm	145.49	Th	0200
Berks Co. ARES/RACES Net	fm	146.91	T	0100
Blue Knob Emergency Preparedness Net, western Penna.	fm	147.15	M	0100
Bradford Co. ARES	fm	147.15	Th	0200
Breakfasteers Net, Beaver Co.	fm	146.85	M-F	1345
Butler Co. Traffic & Weather Net	fm	147.36	Dy	0310
Butler County Public Service Net	fm	147.36	T	0200
Carbon County RACES/ARES Net	fm	147.54	W	2100
Carbon County RACES/ARES Net	fm	224.26	W	2100
Central Counties ARES/RACES Net	fm	146.76	M	0200
Chester County ARES Net	fm	146.94	Th	0030
Crawford County ARES and SKYWARN nets	fm	145.13	on call	
Delaware Co. ARES Net	fm	147.36	Th	0030
Delaware Co. ARES Net	fm	224.50	Th	0030
District 2 ARES Net, SCentrl Pa.	fm	145.45	F	0100
District 5 Emergency Services Net, Monroe Co.	fm	146.865	M	0100
District 5 Emergency Services Net, Pike Co.	fm	146.865	M	0100
District 5 Emergency Services Net, Wayne Co.	fm	146.865	M	0100
District 6 ARES, Susquehanna	fm	147.00	W,F	0100
District 8 ARES, Montour	fm	147.30	W	0000
Eastern Area RACES Net	lsb	3.9875	M,W,F	2200
Eastern Pennsylvania CW Net	cw	3.610	Dy	0000
Eastern Pennsylvania CW Net	cw	3.610	Dy	0300
Eastern Pennsylvania Emergency Phone Traffic Net	lsb	3.917	Dy	2300
Erie Co. ARES Net- SKYWARN	fm	146.61	M	0230
Fayette & Westmoreland Cos. ARES Net	fm	146.67	M	0200
Grandfathers Net	lsb	3.9625	M-F	1330
Huntingdon County ARES Net	fm	146.70	M	0230
Indiana Co. ARES Net	fm	146.91	T	0100
Lake Erie Emergency Net	usb	29.000	M	0100
Lycoming Co.-District 7 ARES Net	fm	146.73	M	2345
Mercer Co. ARES/RACES	fm	147.15	F	0200
Penn Wireless Assn. ARES Net, SE Pa.	fm	146.715	M	0100
Penn-Mar Radio Club, E Pa.	fm	147.33	W	0000
Pennsylvania State Civil Defense RACES Net	lsb	3.9935	1st M	1300
Rescue 40 Search & Disaster Team Net	fm	147.165	Th	0200
Schuylkill Co. Emergency Services Net, Berks Co.	fm	145.37	T	0100
Schuylkill Co. Emergency Services Net, Schuylkill Co.	fm	145.37	T	0100
Somerset County Emergency Services		144.795	Sn	2330
Tioga County Amatuer Radio Club Net	lsb	3.985	Sn	0100
Tri-State Emergency Net, northwest Pennsylvania	fm	146.61	Sn	2230
Triple "A" ARA 2-Meter Net, Beaver Co.	fm	146.85	T	0200
Triple "A" ARA RTTY Net, Beaver Co.	fm	147.135	S	0200

Triple States ARES CW Net, southwest Pennsylvania	usb	28.605	Th	0030
Triple States ARES Net, southwest Pennsylvania	usb	28.605	Th	0100
Triple States Weather Net	fm	146.91	W	0015
Washington County SKYWARN	fm	146.79	on call	
West Virginia Weather & Travelers Net	fm	145.27	T	0000
West Virginia Weather & Travelers Net	fm	147.285	T	0000
West Virginia Weather & Travelers Net	fm	444.400	T	0000
Western Pennsylvania CW Net	cw	3.585	Dy	0000
Western Pennsylvania Phone & Traffic Net	lsb	3.983	Dy	2300
Windjammers, Philadelphia		50.325	M-Sa	1600
York Co. RACES/ARES	fm	146.97	M	0200
Puerto Rico				
PRAC Net	fm	147.09	Sn	1300
Puerto Rico Weather & Emergency Net	lsb	3.9265	Dy	1110
Puerto Rico Weather & Emergency Net	lsb	3.9265	Dy	2310
West Indies Net-Central	fm	147.33	Dy	2230
Rhode Island				
Aquidneck Island Communications Net	fm	147.36	Sn	1315
Bristol Co. ARES Net	fm	145.33	Sn	0130
Kent Co. ARES Net	fm	145.35	Sn	2330
New England Emergency Phone Net	lsb	3.945	Sn	1330
Newport Co. ARES Net	fm	147.36	Sn	1315
Newport Co. Red Cross Net	fm	147.36	1st F	0030
Providence Co. ARES Net	fm	145.17	1st,3rd Th	0000
Providence Co. ARES Net	fm	224.56	1st,3rd Th	0000
Rhode Island ARES Net	fm	146.70	M	0100
Rhode Island ARES/Red Cross Net	fm	146.70	W	0100
Washington Co. ARES Net	fm	147.165	M	0030
South Carolina				
Carolina State Line Net, Centrl Savannah River	fm	146.73	Sn	0200
Coastal Carolina Emergency Net	lsb	3.907	Dy	0000
Edisto ARS 2-Meter Net, Orangeburg	fm	147.09	3rd F	0100
Jumpsuit Net	usb	14.180	M-F	2000
Laurens Co. ARES Net	fm	146.965	Sn	0130
Piedmont Emergency Training Net	fm	145.35	T-M	0100
SC ARES/RACES Net	lsb	3.9935	1st M	2100
SC Noontime Net	lsb	7.243	M-Sa	1700
SC Noontime Net	lsb	3.905 alt.	M-Sa	1700
SC Singlesideband Net	lsb	3.915	Dy	0000
SC Singlesideband Net	lsb	7.243 alt.	Dy	0000
Trident ARC 2-Meter Net, low country	fm	147.27	W	0100
York Co. 2-Meter Net	fm	147.03	T-S	0030
South Dakota				
Dakota Chapter 102 QCWA Net, statewide	lsb	3.890	Sn	1400
Northeastern South Dakota 2-Meter Net, Watertown	fm	146.85	T	0300
Sioux Empire ARC Net, Sioux Falls	fm	146.76	T	0045
South Dakota CW Net	cw	3.650	M-F	0100
South Dakota Net-NEO Session	lsb	3.870	Dy	0000
South Dakota Net-NJQ Session	lsb	3.870	Dy	1815
South Dakota Sunday Emergency Net	lsb	3.960	Sn	1500
Tri-State Emergency Weather Net	fm	146.85	M-F	0400
Walworth Co. Emergency Net	cw	3.740	Sn	1830
Tennessee				
135 Repeater Group, southeast Tennessee	fm	147.135	Th	0200
Bristol, TN ARES Net	fm	146.67	T	0100

Net	Mode	Freq	Days	Time
Chattanooga Area Weather Net	fm	146.61	W	0100
Dalton Ga. Amteur Radio Club, southeast Tennessee	fm	145.23	M	0230
Dalton Ga. ARC ARES, southeast Tennessee	fm	145.23	S	0300
East Tennessee Hospital Net	fm	147.12	W	0100
Middle East Tennessee Emergency Net	fm	147.30	T-S	0200
North Georgia 220 Net, southeast Tennessee	fm	224.74	on call	
Northwest Georgia ARES Net, southeast Tennessee	fm	146.715	S	0300
Tennessee Civil Defense Weather Net	fm	146.91	W	0200
Tri State Two Meter Net	fm	147.30	Sn,T,Th	2130
West Tennessee Weather Net	fm	146.97	Dy	0230

Texas

Net	Mode	Freq	Days	Time
ARES/RACES Cloud Chasers Net, panhandle	fm	146.67	M	0200
ARES/RACES Wind Jammers, panhandle	fm	147.30	T	0200
Bay Town City Emergency Management	fm	146.78	W	0300
Big Bend Emergency Net, SW Texas	lsb	3.922	Sn	1430
Brazos County ARES Net	fm	146.68	1st W	0300
Central Texas Emergency Net	cw	3.760	Sn	1300
Central Texas Emergency Net	lsb	3.910	Sn	1400
Clear Lake Emergency ARS	fm	146.64	M	2359
Cleveland ARES Net	fm	146.90	Th	0230
Crossroads Area Weather Net	fm	144.69	W	0100
Dallas Combined ARES, Dallas and Fort Worth	fm	147.06	M	0200
Dallas RACES	fm	146.88	1st,3rd M	0100
Davis Co. Emergency Net	fm	147.42	F	0200
District 7 Emergency Coordinator Net, Waco	fm	145.55	1st,3rd M	0300
District 33 RACES Net, Lower Rio Grande Valley	fm	146.06	S	0100
El Paso 2-Meter Emergency & Training Net, El Paso	fm	146.88	Th	0200
Fayette County Emergency Management Net	fm	146.20	M	0200
Fayette County Emergency Net	fm	146.80		0200
Fort Bend Co. EMA	fm	147.30	T	0000
Fort Bend Emergency Management Net	fm	147.30	M	0200
Galveston City ARES Net	fm	147.14	W	0100
Galveston City RACES	fm	146.66	W	0200
Garland RACES	fm	147.24	F	0200
Harris County ARES/RACES Net, Gulf Coast	fm	146.94	Sn	0100
Hays County RACES Net	fm	147.10	2nd Th	0030
Henderson Co. ARES Net	fm	147.02	Th	0200
Hill Co ARES Training Net	fm	147.00	F	0200
Houston City Emergency Management Net	fm	147.14	W	0230
Longhorn Net, Abilene	fm	147.81	F	0130
Lower Rio Grande Valley Emergency Net	fm	146.96		2000
McCulloch County Emergency Net	fm	146.90		0300
Mercury Amateur Radio Association	lsb	3.994	S	0230
Montgomery County ARES	fm	146.02	M	0130
Northeast Texas Emergency Net	lsb	3.970	Sn	1400
Northwest Amateur Radio Society ARES, NW Houston	fm	146.66	W	0200
Panhandle Traffic & Emergency Net	lsb	3.933	Dy	0000
Plano ARES	fm	147.18	1st,3rd M	0200
Red River Valley Net, NE Tex.	fm	146.76	W	0300
San Antonio Area Rainfall Net	fm	146.94	Dy	0200
San Benito ARC Emergency Net	fm	146.06	Sn	0200
South Texas Emergency Net - CW Section	cw	3.693	W	0100
South Texas Emergency Net - Zone 1	lsb	3.955	Sn	1330
South Texas Emergency Net - Zone 2	lsb	3.955	W	0015
South Texas Emergency Net - Zone 4	lsb	3.955	Sn	1500

Net	Mode	Freq	Day	Time
South Texas Emergency Net	lsb	3.955	T	0030
South Texas Emergency Net	lsb	3.955	Th	0100
South Texas Emergency Net	lsb	7.248	T	0100
South Texas Emergency Net VHF Section, Victoria	fm	144.59	W	0130
STARS Packet Net, Rio Grande Valley	fm	145.01	W	0200
State RACES Net	lsb	7.248	2nd,4th Sn	1930
TEXAS ARES (daytime)	lsb	7.273	on call	
TEXAS ARES (nighttime)	lsb	3.873	on call	
Texas Ama. Radio Emerg. Service Net	lsb	3.873	Sn	0130
TIERS, Gulf Coast	fm	147.15	Sn	0000
Valleywide Emerg. Net, Lower Rio Grande Valley	fm	147.39	M	0000
Washington Co Emergency Preparedness Net	fm	147.26	M	0200
Weber Co. Emergency Services	fm	146.90	W	0230
Webfoot, Rio Grande Valley	fm	146.76	Dy	1430
West Gulf Emergency Net	lsb	3.9435	Sn	1400
West Gulf Emergency, Galveston and Houston	fm	145.37	Sn	1430
West Houston RACES Net	fm	146.06	S	1800 local
Wharton Co. RACES Net	fm	144.27	T	0100
Wichita Co. ARES	fm	146.94	T	0100

Utah

Net	Mode	Freq	Day	Time
Beehive Utah Net	lsb	7.272	Dy	0230
Color Country ARES Net, Kane Co.	fm	146.88	M	0230
Mercury Amateur Radio Association	lsb	3.983	S	1500
Nevada Weather Net, Utah	lsb	3.993	M-Sa	0600
Utah Code Net	cw	3.710	Dy	0230

Vermont

Net	Mode	Freq	Day	Time
Bennington Co. ARES Net	fm	145.39	T	0130
Bennington Co. ARES Net	usb	28.333	T	0100
Connecticut Valley FM Association Weather Net	fm	146.76	M-F	1130
Connecticut Valley FM Association Weather Net	fm	146.76	M-F	2200
Connecticut Valley FM Association Weather Net	fm	146.76	T-S	0300
Hassall's Weather Net	fm	146.745	Dy	0200
Northern Vermont ARES Training Net	fm	145.47	2nd,4th T	0100
Tri-State Emergency Net	fm	146.805	Th	0030
Twin State Emergency Net	fm	146.76	W	0030
Vermont Phone Emergency Net	lsb	3.934	Sn	1530
Vermont Sideband Net	lsb	3.909	M-Sa	2300
Vermont Sideband Net	lsb	3.909	Sn	1300

Virginia

Net	Mode	Freq	Day	Time
Allegheny Regional Emergency Service Net	fm	146.805	Dy	0100
Arlington Co. ARES Net	fm	145.47	on call	
Arlington Co. ARES Net	fm	147.045	W	0030
Bristol ARES	fm	146.67	W	0130
Chesapeake ARES Net, south Tidewater	fm	146.61	T	0100
Chesterfield ARES Net	fm	146.94	M	0200
Clinch Valley Emergency Net	fm	146.835	T	0230
Eastern Shore ARES Net, Accomack Co.	fm	147.255	T	0200
Eastern Shore ARES Net, Northhampton Co.	fm	147.255	T	0200
Frederick County ARES Net, Winchester	fm	146.82	Sn	2030
Gloucester ARES	fm	145.37	F	0100
Hampton ARES	fm	146.73	F	0100
Hanover County ARES Net	fm	145.43	T	0030
Henrico ARES Net	fm	147.51	F	0030
Hopewell-Prince George ARES	fm	147.39	M	0030
Metropolitan Rptr Assn Liaison ARES Net, Richmond	fm	145.43	M	0200

Net	Mode	Freq	Day	Time
Mt. Vernon ARES Net	fm	146.655	W	0100
National Capital ARES, Washington DC	fm	146.91	M	0200
New Kent ARES	fm	147.33	Th	0030
Newport News ARES	fm	147.165	W	0100
Northern Piedmont Emerg. Net, Charlottesville	fm	146.76	T	0030
Ole Virginia Hams Bulletin Net, northern Virginia	fm	146.97	F	0100
Portsmouth ARC Net	fm	146.85	W	0100
Richmond ARES Net	fm	145.11	W	0100
Roanoke Emergency Service Net	fm	146.745	W	0030
Rockingham ARES Net	fm	147.225	M	2300
Shenandoah Valley Emergency Net	fm	146.82	Dy	0015
Smyth Co. ARES Net	fm	146.64	F	0200
South Tidewater ARES	fm	146.97	Dy	0200
Triple States ARES High-Speed CW Net	cw	28.480	Th	0010
Triple States ARES Intermediate CW Net	cw	28.480	W	2350
Triple States ARES Phone Net	usb	28.480	Th	0030
Tug Valley ARES/RACES Net	fm	145.33	M	0200
Vienna Wireless Society Emergency Net	fm	146.685	F	0030
Virginia ARES & ARRL Appointee Net	lsb	3.947	3rd W	0130
Virginia Emergency Net, Alpha	lsb	3.910	on call	
Virginia Emergency Net, Bravo	lsb	3.947	on call	
Virginia Emergency Net, Charlie	cw	3.680	on call	
Virginia Late Net	lsb	3.947	Dy	0315
Virginia Net-Early	cw	3.680	Dy	0000
Virginia Net-Late	cw	3.680	Dy	0300
Virginia Sideband Net	lsb	3.947	Dy	2300
Virginia Slow Net	cw	3.680	Dy	2330
Virginia Traffic Net	lsb	3.907	Dy	1300
Virginia Traffic Net	lsb	7.260 alt.	Dy	1300
Washington Co. ARES Net	fm	146.64	F	0200
West Virginia Weather & Travelers Net	fm	145.27	T	0000
West Virginia Weather & Travelers Net	fm	147.285	T	0000
West Virginia Weather & Travelers Net	fm	444.40	T	0000
Williamsburg Area ARES Net	fm	146.67	T	0045
WVA, MD, VA Net	lsb	3.905	M-Sa	2200
York Co. Poquoson ARES	fm	146.94	T	0100

Virgin Islands

Net	Mode	Freq	Day	Time
Caribbean Maritime Mobile Net	lsb	7.237	Dy	1100
St Croix ARES	fm	147.250	Sn	2145
St Thomas/St. John ARES Net	fm	146.810	Sn	2200
Virgin Island ARES Net	fm	146.81	Sn	2300
Virgin Island ARES Net	fm	147.25	Sn	2300
Virgin Islands Net East	ssb	1.984	Dy	0001

Washington

Net	Mode	Freq	Day	Time
Gilmer Co. ARES Net	fm	145.29	F	0400
King County ARES Net	fm	145.33	M	0400
Kitsap Emergency ARES/RACES Net	fm	145.31	M	0300
Kitsap Emergency ARES/RACES Net; Leadership Net	lsb	3.987	M	0300
Mason County ARES	fm	146.72	Th	0330
Nevada Weather Net, Washington	lsb	3.993	M-Sa	0600
Panhandle Emergency Services, Spokane	fm	147.78	W	0330
Pierce Co. ARES	fm	147.38	W	0300
Snohomish Co. ARES	fm	147.78	Sn	0300
Snohomish Co. ARES	fm	147.78	W	0330
South Snohomish Co. ARES Info Net	fm	146.18	M	0400

Spokane ARES Net	fm	147.30	W	0400
Team Bravo ARES Net, East King Co.	fm	145.33	W	0400
Team Charlies ARES Net, West King Co.	fm	145.33	T	0400
Team Echo/Tango ARES Net, South King Co.	fm	145.33	Th	0300
Team Victor ARES Net, Vashon Island, King Co.	fm	147.48	F	0400
Walla Walla County ARES	fm	147.88	F	0400
Washington Amateur Radio Traffic System, summer	lsb	3.970	Dy	0100
Washington Amateur Radio Traffic System, winter	lsb	3.970	Dy	0200
Washington Emergency Net	lsb	3.987	S	1700
Washington Emergency Net	lsb	3.987	T	0230
Whatcom Co. ARES	fm	145.23	T	0130
Whatcom Co. ARES	fm	147.06	M	0400
Whatcom Co. ARES	fm	147.16	M	0330
Whatcom Co. ARES	lsb	3.965	T	0145

West Virginia

Harrison Co. ARES/RACES Net	fm	146.685	F	0030
Marion County ARES/Weather Net	fm	146.90	T	0200
Marion County ARES/Weather Net	usb	28.400	T	0130
Marshall Co. ARES Net	fm	146.91	T	0110
Marshall Co. ARES Net	usb	28.480	T	0100
Monongalia County ARES/Weather Net	fm	146.90	T	0200
Northern Panhandle/Wheeling Emerg. Net	fm	146.76	T	0115
Ohio County ARES Net	fm	145.19	M	0100
Pendelton Co. ARES/RACES	fm	147.285	F	0000
Taylor County ARES/Weather Net	fm	146.90	T	0200
Taylor County ARES/Weather Net	usb	28.400	T	0130
Triple States ARES High-Speed CW Net	cw	28.480	Th	0010
Triple States ARES Intermediate CW Net	cw	28.480	W	2350
Triple States ARES Phone Net	usb	28.480	Th	0030
Triple States Weather Net SKYWARN	fm	146.91	W	0015
Tug Valley ARES/RACES Net	fm	145.330	M	0200
WC8AAK ARES/RACES Net, Brooke	fm	146.94	M	0100
WC8AAK ARES/RACES Net, Hancock	fm	146.94	M	0100
West Virginia Amateur Radio ARES-Info, Charleston	fm	147.27	M	0030
West Virginia Amateur Radio ARES-Info, Huntington	fm	147.27	M	0030
West Virginia ARES/RACES Net	lsb	3.865	Th	2330
West Virginia Early Net	cw	3.567	Dy	0000
West Virginia Fone Net	lsb	3.865	Dy	2300
West Virginia Weather & Travelers Net	fm	145.27	T	0000
West Virginia Weather & Travelers Net	fm	147.285	T	0000
West Virginia Weather & Travelers Net	fm	444.40	T	0000
WVA, MD, VA Net	lsb	3.905	M-Sa	2200

Wisconsin

ARES Net, southeast Wisconsin	fm	145.25	T	0200
Badger Emergency Net	lsb	3.985	Dy	1800
Badger Weather Net	lsb	3.985	M-Sa	1200
ChicagoLink repeater system, southeast Wisconsin	fm	29.680	M	2330
ChicagoLink repeater system, southeast Wisconsin	fm	52.825	M	2330
ChicagoLink repeater system, southeast Wisconsin	fm	145.45	M	2330
ChicagoLink repeater system, southeast Wisconsin	fm	147.075	M	2330
ChicagoLink repeater system, southeast Wisconsin	fm	147.225	M	2330
ChicagoLink repeater system, southeast Wisconsin	fm	223.90	M	2330
ChicagoLink repeater system, southeast Wisconsin	fm	224.16	M	2330
ChicagoLink repeater system, southeast Wisconsin	fm	224.48	M	2330
ChicagoLink repeater system, southeast Wisconsin	fm	224.80	M	2330

ChicagoLink repeater system, southeast Wisconsin	fm	444.75	M	2330
Milwaukee ARES Net	fm	146.67	T	0300
Milwaukee-Florida Net	usb	14.290	Dy	1400
North Central Phone Net	lsb	3.913	M-Sa	13+1800
Outagamie County ARES	fm	146.76	Th	0300
Portage County ARES	fm	146.985	T	0300
RACES, Statewide	lsb	3.9935	Sn	1300
RACES, Statewide	lsb	3.9935	Th	1715
Sheboygan Co. ARES	fm	147.66	T	0100
Sheboygan Co. ARES	lsb	3.860	Sn	1630
Wisconsin Sideband Net	lsb	3.985	Dy	2330

Wyoming

Albany ARES	fm	146.061	M	0145
Area 4 ARES, northwest Wyoming	fm	146.285	Sa,Sn	0145
Area 5 ARES, southwest Wyoming	fm	146.61	Sn	0100
Campbell ARES, Gillette	fm	146.97	S	1600
Casper ARES	fm	146.94	S	0300
N7JJY packet bulletin board system	pkt	145.01	24 hours	
Sheridan ARES	fm	146.82	T	0215
Wyoming ARES/RACES Net	lsb	3.923	Sn	1600
Wyoming Cowboy/Pony Express/ARES	lsb	3.923	M-F	0045
Wyoming Jackalope Net	lsb	7.260	M-Sa	1915
Wyoming Pony Express Net	lsb	3.923	Sn	1500

U.S. Nationwide And Regional Nets

U.S., Nationwide

Buck-A-Roo Net	lsb	7.2645	M-F	1100
Disciples Amateur Radio Fellowship	lsb	3.883	M-Sa	1230
Disciples Amateur Radio Fellowship	lsb	7.270	M,W,F	1830
Disciples Amateur Radio Fellowship	usb	14.287	T,S	1530
Mercury Amateur Radio Association	lsb	7.294	M	0800
Mercury Amateur Radio Association	usb	28.350	Sn	1915
Seven Day Weekenders Net	usb	14.242	M-F	2030
Ten Meter Youth Net	usb	28.375	W	2100
Veteran Administration Amateur Radio System	usb	14.287	M-F	1700
Winnebago-Itasca Travel Club	usb	21.393	S	1830

U.S., East

Central Gulf Coast Hurricane Net, U.S. southeast	lsb	3.935	Dy	0100
Mercury Amateur Radio Association, U.S. northeast	lsb	3.913	T	0245
Mercury Amateur Radio Association, U.S. northeast	lsb	7.228	T	0315
Mercury Amateur Radio Association, U.S. southeast	lsb	3.875	S	1100
Mercury Amateur Radio Association, U.S. southeast	lsb	3.875	Th	0200
Nerd Net, U.S. east	lsb	3.868	Dy	0200
New England Emergency Phone Net, U.S. New England	lsb	3.945	Sn	1330
New England Weather Net, U.S. East Coast	lsb	3.905	M-Sa	1030
Old Goats Net, U.S. east	lsb	7.210	Dy	0030
Recreational Vehicle Service Net, U.S. east	lsb	7.233	Dy	1200
Recreational Vehicle Service Net, U.S. northeast	lsb	3.895	M-F	2200
Recreational Vehicle Service Net, U.S. northeast	lsb	3.963	Sn	1300
Seacoast Net, U.S. New England	lsb	3.895	Sn	1500
South Coast Amateur Radio Service Net, U.S. southeast	lsb	7.251	Dy	1200
Ten Meter East Coast Hurricane Net, U.S. East Coast	usb	28.313	Dy	1500
Vermont Sideband Net, U.S. northeast	lsb	3.909	M-Sa	2300
Vermont Sideband Net, U.S. northeast	lsb	3.909	Sn	1300
Waterway Net, U.S. East Coast	lsb	7.268	Dy	1145

Waterway Radio & Cruising Club Net, U.S. East Coast	lsb	7.268	Dy	1245

U.S., Central

7290 Traffic Net, U.S. south central	lsb	7.290	M-F	1900
7290 Traffic Net, U.S. south central	lsb	7.290	M-Sa	1600
Central Gulf Coast Hurricane Net, U.S. Gulf Coast	lsb	3.935	Dy	0100
Episcopal Clergy Roundtable, U.S. Great Lakes	lsb	7.290	T	1800
Great Lakes Emergency & Traffic Net, U.S. Great Lakes	lsb	3.932	Dy	0200
Magnolia Net, U.S. central	lsb	3.877	Dy	0030
Magnolia Net, U.S. central	lsb	3.877	Dy	1200
Mercury Amateur Radio Association, U.S. Midwest	cw	3.705	S	1230
Mercury Amateur Radio Association, U.S. Midwest	lsb	3.875	Th	0200
Mercury Amateur Radio Association, U.S. Midwest	lsb	3.918	S	1600
North Central Amateur Radio Service, U.S. north central	lsb	7.241	M-F	2200
North Central Amateur Radio Service, U.S. north central	lsb	7.245	M-F	1400
Recreational Vehicle Service Net, U.S. central	lsb	7.233	Dy	1200
Salvation Army Team Emergency Net, U.S. Midwest	lsb	7.265	S	1530
West Gulf Hurricane Net, U.S. Gulf Coast	lsb	7.268	T	1805

U.S., West

California Traffic Net, U.S. west	lsb	3.905	Dy	0230
California/Hawaiian Net, maritime service, Pacific Coast	usb	14.340	M-F	1630
Chubasco Net, maritime service, U.S. Pacific Coast	lsb	7.294	Dy	1530
Disciples Amateur Radio Fellowship, U.S. west	lsb	3.940	M-Sa	1500
Mercury Amateur Radio Association, U.S. northwest	lsb	3.870	Th	0300
Mercury Amateur Radio Association, U.S. southwest	lsb	7.228	F	0300
Mercury Amateur Radio Association, U.S. west	cw	3.625	Th	0330
Mercury Amateur Radio Association, U.S. west	lsb	3.985	W	0330
Mercury Amateur Radio Association, U.S. west	lsb	7.228	S	1500
Nitwits Over The Hill Inspirational Nut Group, U.S. west	lsb	3.933	Dy	1400
Pacific Coast Net, U.S. Pacific Coast	cw	7.085	Dy	1900
Recreational Vehicle Service Net, Rocky Mountain states	lsb	7.263	M-F	1500
Recreational Vehicle Service Net, U.S. Pacific Coast	lsb	7.263	M-F	1600
West Coast Admirals MM Net, maritime, Pacific Coast	lsb	7.190	Dy	2230
West Coast Amateur Radio Service, U.S. west	lsb	7.255	Dy	2000
Western Public Service System, U.S. west	lsb	3.952	Dy	0230

Canadian And Mexican Nets

Maritime Provinces

Atlantic Provinces Net	cw	3.654	Dy	0000

Newfoundland

Atlantic Provinces Net	cw	3.654	Dy	0000

Quebec

Emegency Auxiliary Net Quebec Civil Protection Bureau	lsb	3.760	1st W	0030
Emegency Auxiliary Net Quebec Civil Protection Bureau	lsb	7.060	1st W	0030
Happy Gang, provincewide	lsb	3.765	Dy	1300
Le P'tit Train du Matin, provincewide	lsb	3.750	Dy	1300
Le Reseau de la Detente, provincewide	lsb	3.750	Dy	2200
Le Reseau du Quebec, provincewide	lsb	3.780	Dy	2345
Quebec Radio Net	lsb	3.775	Dy	0030
Western Quebec VHF/UHF ARES Net	fm	147.00	Dy	0030

Ontario

ARES Kingston	fm	146.94	W	0015
ARES Net Canada, nationwide	usb	14.115	Sn	2000
ARES Net, provincewide	cw	7.060	1st Sn	1900
Champlain Mini Net	lsb	147.06	Dy	2345
Communications Ontario		7.074	Dy	1500

Eastern Cos. & Cornwall ARES Net	fm	147.225	W	0130
Huronia Amateur Radio Emergency Services Net	fm	147.15	T	2000
Laurentian Net, provincewide	lsb	3.755	Dy	2345
Nightly 2 Meter ARES Net, Sudbury	fm	147.06	Dy	0200
Ontario Amateur Radio Service	lsb	3.755	Dy	1200
Ontario Phone Net	lsb	3.770	Dy	0000
Transprovincial Net		7.055	Dy	1500
York Region ARES Net	fm	147.225	Sn	0100

Manitoba

Manitoba Evening Phone Net	lsb	3.759	Dy	0100
Manitoba Morning Net	lsb	3.743	Dy	1300
Manitoba Traffic Net	cw	3.660	Dy	0030

Saskatchewan

Prairie Weather Net	lsb	3.780	Dy	1430
Saskatchewan Amateur Traffic Net	cw	3.695	Dy	0130
Saskatchewan ARES Net	lsb	3.780	Sn	1530
Saskatchewan Phone Net	lsb	3.753	Dy	0100

Alberta

Alberta Radio Emergency Service	lsb	3.750	Sn	1525
Calgary Amateur Radio Emergency Net	fm	146.85	F	0330
Wee Net, provincewide	lsb	3.770	Sn	1600

British Columbia

British Columbia Emergency Net		3.650	Dy	0300
British Columbia Public Service Corp.	lsb	3.729	Dy	0130

Yukon

The Alaska Bush Net	lsb	7.091	Dy	0500

Canadian Regional Nets

Canadian ARES Net, nationwide	usb	14.130	Sn	2000
Disciples Amateur Radio Fellowship, nationwide	lsb	3.883	M-Sa	1230
Disciples Amateur Radio Fellowship, nationwide	lsb	3.940	M-Sa	1500
Disciples Amateur Radio Fellowship, nationwide	usb	14.287	T,S	1530
Maritime Emergency Net, northeast Canada	usb	14.310	Dy	0400
Maritime Weather Net, northeast Canada	lsb	3.750	Dy	1100
Mercury Amateur Radio Association, western Canada		3.625	Th	0330
Mercury Amateur Radio Association, western Canada	lsb	3.870	Th	0300
Mercury Amateur Radio Association, western Canada	lsb	7.228	S	1500
Nerd Net, eastern Canada	lsb	3.868	Dy	0200

Mexico

Baja California Maritime Net	lsb	7.2385	Dy	1600
Chubasco Net, west coast, maritime service	lsb	7.294	Dy	1530
Mexican Emergency Net, Sonora	lsb	7.090	Dy	0300
SONRISA, west coast, maritime service	lsb	3.968	Dy	1400

Western Hemisphere Nets

14295 Coordinators Net, North America	usb	14.295	Dy	1300
Disciples Amateur Radio Fellowship, Westrn Hemisphere	usb	21.415	T	2100
Hurricane Watch Net, Western Hemisphere	usb	14.325	Dy	
Panamerican Emergency Net, North & Central America	usb	14.130	Dy	2300
Recreational Vehicle Service Net, North America	usb	14.308	M-F	1700
Recreational Vehicle Service Net, North America	usb	14.308	M-F	2200

International Nets

Worldwide

Intercontinental Traffic Net, worldwide	usb	14.313	Dy	1000

International Assistance and Traffic Net, worldwide	usb	14.303	Dy	1100
International Association of Airline Hams, worldwide	usb	14.280	Sn,W	1500
International Association of Airline Hams, worldwide	usb	21.380	Th	1500
International CW Traffic Net, worldwide	cw	14.040	Dy	1230
International CW Traffic Net, worldwide	cw	21.040	Dy	2230
Mercury Amateur Radio Association, Europe	lsb	3.913	T	0245
Mercury Amateur Radio Association, Europe	lsb	7.228	T	0315
Mercury Amateur Radio Association, Oceania	lsb	7.294	M	0800
Mercury Amateur Radio Association, worldwide	usb	14.287	F	0415
Mercury Amateur Radio Association, worldwide	usb	14.287	S	0600
Radio Amateur's Emerg. Net, East Sussex Co., England	fm	144.85	F	1900
Southeast Asia Net, worldwide	usb	14.320	Dy	1200
Three Blind Mice Net, worldwide	usb	14.270	Dy	0330

Maritime nets, Worldwide

Atlantic Maritime Net, Atlantic Ocean	usb	14.303	Dy	0530
Atlantic Maritime Net, Atlantic Ocean	usb	14.303	Dy	1630
Atlantic Maritime Net, Atlantic Ocean	usb	14.303	Dy	2030
Atlantic Maritime Net, Indian Ocean	usb	14.303	Dy	0530
Atlantic Maritime Net, Indian Ocean	usb	14.303	Dy	1630
Atlantic Maritime Net, Indian Ocean	usb	14.303	Dy	2030
Atlantic Maritime Net, Pacific Ocean	usb	14.303	Dy	0530
Atlantic Maritime Net, Pacific Ocean	usb	14.303	Dy	1630
Atlantic Maritime Net, Pacific Ocean	usb	14.303	Dy	2030
Baja California Maritime Net, Sea of Cortez	lsb	7.2385	Dy	1600
California to Caribbean	usb	14.285	M	2300
California to South Pacific, maritime service net	usb	14.285	M	2310
Caribbean Maritime Mobile Net	lsb	7.230	Dy	1100
Caribbean Net	lsb	7.158	Dy	0000
Confusion Net, maritime service net, Pacific Ocean	usb	14.305	M-F	1900
DDD Net-Pacific for Canadians, maritime service net	usb	14.115	Dy	0400
DDD Net-Pacific for Canadians, maritime service net	usb	14.115	M-F	1730
International Maritime Mobile Net, worldwide	usb	14.313	Dy	0000
Mariana Net, maritime service net, Pacific Ocean	usb	14.340	M-Sa	1900
Mariana-Guam, maritime service net, Pacific Ocean	usb	14.310	Dy	0700
Maritime Mobile Net, worldwide	usb	28.380	Dy	1200
Maritime Mobile Service Net, worldwide	usb	14.313	Dy	1800
Mississauga Maritime Mobile Net, Caribbean Sea	usb	14.121	Dy	1145
Mississauga Maritime Mobile Net, South Atlantic	usb	14.121	Dy	1145
Pacific Maritime Mobile Net	usb	14.314	Dy	0400
Pacific Maritime Mobile Net	usb	21.404	M-F	2230
Persian Gulf Rag Chew Net	usb	28.380	S,Sn	1200
SONRISA, maritime service, Sea of Cortez	lsb	3.968	Dy	1400
South Pacific Net, maritime service net	lsb	3.815	Dy	0715
South Pacific Net, maritime service net	usb	14.315	Dy	0800
Southern California 10 Meter Maritme	usb	28.313	Dy	0130
Ten Meter Maritime Mobile Net, Pacific Rim	usb	28.480	Dy	0300
Trans Atlantic Maritime Mobile, Caribbean Sea	usb	21.400	Dy	1300
Trans Atlantic Maritime Mobile, North Atlantic	usb	21.400	Dy	1300
Trans Atlantic Maritime Mobile, South Atlantic	usb	21.400	Dy	1300
United Kingdom Net, Atlantic Ocean	usb	14.303	Dy	0800
United Kingdom Net, Atlantic Ocean	usb	14.303	Dy	1800
United Kingdom Net, maritime service net, Pacific Ocean	usb	14.303	Dy	0800
United Kingdom Net, maritime service net, Pacific Ocean	usb	14.303	Dy	1800
Waterway Radio & Cruising Club Net, Atlantic Ocean	lsb	7.268	Dy	1245
Waterway Radio & Cruising Club Net, Caribbean Sea	lsb	7.268	Dy	1245

SWL Nets For Frequency Updates

There are amateur radio nets relaying the latest shortwave listening news and DX tips, including changes in weather radio frequencies. Best known is a net conducted by ham members of the Association of North American Radio Clubs (ANARC). Listeners call in tips, using their own ham stations or by telephoning a ham who repeats the info on the air.

Net	Mode	MHz	Day	UTC
ANARC SWL Ham Net, eastern North America	lsb	7.240	S	1000 Eastern
Great Circle Shortwave Society Net, nationwide	lsb	3.855	M	0100
The Shortwave Listener's Net, central New York	fm	145.49	S	0100
The Shortwave Listener's Net, central New York	fm	147.105	S	0100

W1AW Special Emergency Bulletins

The American Radio Relay League (ARRL) headquarters station W1AW transmits special bulletins during large-scale communications emergencies on the frequencies below:

Mode	Time	MHz
lsb	on the hour	1.890, 3.990, 7.290
usb	on the hour	14.290, 21.390, 28.590
ssb	on the hour	50.190
fm	on the hour	147.555
cw	on the half hour	1.818, 3.5815, 7.0475, 14.0475, 21.0775, 28.0775, 50.080, 147.555
rtty	15 min. past the hour	3.625, 7.095, 14.095, 21.095, 28.095, 147.555

Wind Profilers

NOAA plans to station wind-profiling radars on amateur radio frequencies around 449 MHz where hams regularly communicate in the narrow-band fm (fm or nbfm) voice mode.

A wind profiler is a Doppler radar with antenna pointed up to measure wind direction and speed above its location, from 1,500 ft. to 53,000 ft. (10 miles) above the ground. Storms are moved by winds in that range. Commercial aircraft operate in the same altitude range. Wind profilers around 400 MHz detect wind shear above 1,500 ft., but not lower level wind shear near the ground. Wind profilers at 1,000 MHz detect very low wind shear just above the ground where it is a hazard for planes landing on airport runways.

Meteorologists use profiler data to predict atmosphere events such as thunderstorms. Using the information, pilots burn less fuel and fly more safely. Radars also show pollutants, and natural phenomena such as volcanic ash, scattering through the atmosphere.

Frequencies. A profiler antenna points straight up. The altitude its signal reaches depends upon frequency. NOAA needs radars around 400 MHz for the altitude it wants. NOAA already has 31 on the air at 404 MHz—placed away from airports, major highways and metropolitan areas in Connecticut and the midwest from Arkansas to Colorado to Texas to Nebraska—with 200 additional stations on 449 MHz to be built across the country.

Radars shooting 404 MHz signals into the sky interfere with search and rescue satellites listening for weak signals on 406 MHz, so 404 MHz radars will be converted to 449 around 1993-94. Many ham repeaters have outputs around 449. They may not be affected, but repeaters with inputs around 449 are likely to receive interference.

The federal Interdepartmental Radio Advisory System (IRAC) helps the National Telecommunications and Information Administration (NTIA) assign government frequencies. Government radiolocation services, such as radar, have primary use of most UHF bands, while the amateur radio service is allocated secondary use of 420-450 MHz. NOAA also wants wind-profiler radars operating around 50 and 1000 MHz.

Weather Frequencies

The spectrum of electromagnetic energy encompasses all the well-known kinds of radiation, such as radio, television, radar, microwave, X-rays, gamma rays, and even visible, infrared and ultraviolet light.

Places along the spectrum are known as frequencies. That is, one location on the spectrum is a frequency. The electromagnetic spectrum encompasses all frequencies.

Frequency is measured in cycles per second and labeled hertz. One hertz equals one cycle per second. One thousand hertz equals one kilohertz (KHz). One million hertz equals one megahertz (MHz). One billion hertz equals one gigahertz.

The energy we call radio spans a group of frequencies ranging from around 100 hertz to billions of hertz. Radar and microwaves are radio signals found at frequencies of hundreds of millions of hertz. Visible, infrared and ultraviolet light are measured in many billions of hertz and are not radio.

A contiguous group of frequencies within the spectrum is referred to as a band. Bands of frequencies have been labeled with various names, including high frequency (hf), very high frequency (vhf), ultra high frequency (uhf), super high frequency (shf), extremely high frequency (ehf), etc. Smaller segments of the electromagnetic spectrum also have been labeled with letters, such as L-band, S-band, C-band, X-band, Ku-band, K-band, Ka-band, etc. HF also is known as shortwave. Most weather radio signals will be found in the long-wave, medium-wave, shortwave, vhf, uhf and shf regions of the electromagnetic spectrum.

While shortwave signals skip or bounce around the globe, signals on vhf, uhf and higher frequencies are limited to line of sight distances. They don't skip past the horizon.

See the abbreviations explanations at the end of this list for the exact frequencies of various bands and for explanations of abbreviations.

Master List Of Weather Frequencies

MHz	Use
0.060	standard time, WWVB, Fort Collins, Colorado, standard time, operates 24 hours, data
0.050	standard time, OMA, Prague, Czechoslovakia, operates 24 hours
0.050	standard time, RTZ, Irkutsk, Russia, operates 0100-2400 utc
0.060	standard time, MSF, Rugby, Great Britain, operates 24 hrs
0.06666	standard time, RBU, Moscow, Russia, operates 24 hours
0.075	standard time, HBG, Prangins, Switzerland, operates 24 hours
0.0775	standard time, DCF77, Mainflingen, Germany, operates 24 hours
0.162	standard time, Allouis, France, operates 24 hours except Tuesdays 0100-0500 utc
0.198	standard time, RW-166, Irkutsk, Russia, operates 2200-2100 utc
0.272	standard time, RW-76, Novosibirsk, Russia, operates 24 hours
0.418	standard time, ZSC, Capetown, South Africa, 5-min transmission at 0755, 1655 utc
0.434	standard time, VWC, Calcutta, India, 5-minute transmission at 0825 and 1625 utc
0.435	standard time, PPR, Rio de Janeiro, Brazil, 5-min transm'n at 0125, 1425, 2125 utc
0.482	standard time, 4PB, Colombo, Sri Lanka, operates 0553-0600, 1323-1330 utc
0.500	international maritime cw and mcw distress and calling

WEATHER FREQUENCIES

0.500	standard time, PPR, Rio de Janeiro, Brazil, 5-min transm'n at 0125, 1425, 2125 utc
0.500	standard time, VPS, Kowloon, Hong Kong, operates every even hour
0.518	NAVTEX maritime bulletin stations, worldwide, rtty/sitor
0.518	U.S. Coast Guard iceberg warnings, Boston stn NIK, 0445,1045,1645,2245z, navtex
0.530	NOAA Weather Radio relayed by some local Traveler's Information Service stations
0.540-1.600	broadcast stations carrying weather and marine forecasts and warnings, am
1.510	standard time, HD210a, Guayaquil, Ecuador, operates 24 hours
1.610	NOAA Weather Radio relayed by some local Traveler's Information Service stations
1.7505	U.S. Coast Guard disaster communications, channel 1, cw
1.7515	U.S. Coast Guard disaster communications, channel 2, cw
1.7525	U.S. Coast Guard disaster communications, channel 3, cw
1.7535	U.S. Coast Guard disaster communications, channel 4, cw
1.7545	U.S. Coast Guard disaster communications, channel 5, cw
1.7555	U.S. Coast Guard disaster communications, channel 6, cw
1.7565	U.S. Coast Guard disaster communications, channel 7, cw
1.7575	U.S. Coast Guard disaster communications, channel 8, cw
1.7615	U.S. Coast Guard disaster communications, channel 9, usb and cw
1.7685	U.S. Coast Guard disaster communications, channel 10, usb
1.7755	U.S. Coast Guard disaster communications, channel 11, usb
1.7825	U.S. Coast Guard disaster communications, channel 12, usb
1.7895	U.S. Coast Guard disaster communications, channel 13, usb
1.7965	U.S. Coast Guard disaster communications, channel 14, usb
1.800-2.000	amateur radio 160-meter mw band, local, regional, national and international comms
1.818	bulletins on the half hour in major comms emergencies, ARRL hq stn W1AW, cw
1.890	bulletins on the hour in major comms emergencies, ARRL hq stn W1AW, lsb
2.122	WEFAX, Hawaii at 0600-1600x
2.182	international maritime distress frequency, usb
2.195	NOAA weather charts, WEFAX, nationwide, to 0600, usb
2.300-2.498	120-meter worldwide shortwave broadcasting band, tropical band, am
2.400	weather broadcasts by maritime public coast station WMH, Baltimore, Maryland, usb
2.442	weather broadcasts by maritime public coast station WDR, Miami, Florida, usb
2.450	weather broadcasts by maritime public coast stations WAE/WGB, Norfolk, Va., usb
2.450	weather broadcasts by maritime public coast stn KLH, San Francisco, Calif., usb
2.450	weather broadcasts by maritime public coast stn WOU, Boston, Massachusetts, usb
2.466	weather broadcasts by maritime public coast station WFA, Tampa, Florida, usb
2.482	weather broadcasts by maritime public coast station, KOW, Seattle, Washington, usb
2.482	weather broadcasts by maritime public coast stn WAK, New Orleans, Louisiana, usb
2.482	weather broadcasts by maritime public coast stn WQX, New York New York, usb
2.490	weather broadcasts by maritime public coast station WDR, Miami, Florida, usb
2.500	standard time, JJY, Tokyo, Japan, standard time, operates 24 hours, am
2.500	standard time, RCH, Tashkent, Uzbekistan, operates 0500-0400 utc
2.500	standard time, WWV, Fort Collins, Colo., wx, navigation, solar rpts, 24 hrs, am
2.500	standard time, WWVH, Kauai, Hawaii, weather, navigation, solar rpts, 24 hrs, am
2.506	weather broadcasts by maritime public coast stn KLH, San Francisco, Calif., usb
2.506	weather broadcasts by maritime public coast stn WOU, Boston, Massachusetts, usb
2.514	weather broadcasts by maritime public coast station WDR, Miami, Florida, usb
2.522	weather broadcasts by maritime public coast station, KOW, Seattle, Washington, usb
2.522	weather broadcasts by maritime public coast stn WQX, New York New York, usb
2.530	weather broadcasts by maritime public coast station KQP, Galveston, Texas, usb
2.538	weather broadcasts by maritime public coast station WAE, Norfolk, Virginia, usb
2.538	weather broadcasts by maritime public coast station WGB, Norfolk, Virginia, usb
2.538	weather broadcasts by maritime public coast stn KCC, Corpus Christi, Texas, usb
2.550	weather broadcasts by maritime public coast station WFA, Tampa, Florida, usb
2.566	weather broadcasts by maritime public coast station WNJ, Jacksonville, Florida, usb
2.566	weather broadcasts by maritime public coast stn WOU, Boston, Massachusetts, usb

2.572	weather broadcasts by maritime public coast station WLO, Mobile, Alabama, usb
2.590	weather broadcasts by maritime public coast stn WQX, New York New York, usb
2.598	weather broadcasts by maritime public coast stn WAK, New Orleans, Louisiana, usb
2.613	NOAA, Woods Hole, Maine, to National Marine Fisheries Service ships, usb
2.6185	WEFAX, Bracknell, Great Britain, at 1800-0600z, April 1-September 30
2.670	U.S. Coast Guard marine weather broadcast nationwide, emergency coordination, usb
2.704	U.S. Coast Guard emergency command net, rtty and cw
2.81385	WEFAX, Northwood, Great Britain, at 1630-0730z, September 30-March 31
2.815	WEFAX, Moscow at 1800-0055z
2.860	North African region VOLMET weather data for international airline pilots, usb
2.860	West African region VOLMET weather data for international airline pilots, usb
2.863	Pacific Ocean region VOLMET weather data for international airline pilots, usb
2.881	Indian Ocean region VOLMET weather data for international airline pilots, usb
2.905	Atlantic Ocean region VOLMET weather data for international airline pilots, usb
2.905	North Atlantic region VOLMET weather data for internat'l airline pilots, night, usb
2.905	Shannon Aeradio VOLMET weather data for international airline pilots, night, usb
2.956	Middle East region VOLMET weather data for international airline pilots, usb
2.965	Southeast Asian region VOLMET weather data for international airline pilots, usb
2.998	European region VOLMET weather data for international airline pilots, usb
3.0235	international search-and-rescue on-scene, primary frequency, usb
3.046	Halifax military aviation weather, Canada MACS VOLMET, usb
3.1995	Federal Highway Admin. net, nationwide Fed Hiwy Emerg Sys, ch. F1, usb & data
3.200-3.400	90-meter worldwide shortwave broadcasting band, tropical band, am
3.2075	U.S. Coast Guard emergency command net, rtty and cw
3.2081	U.S. Coast Guard emergency command net, usb
3.231	U.S. Air Force Offutt Air Force Base weather, usb WEFAX
3.235	WEFAX, Frobisher Bay CA at 1000-2200z, July 1-October 15
3.2895	WEFAX, Bracknell, Great Britain
3.303	U.S. Dept. of Transportation emergency net, nationwide, channel F1, usb and data
3.3045	Fed. Highway Admin. net, nationwide Fed. Hiway Emerg. Sys., ch. F10, usb & data
3.330	standard time, CHU, Ottawa, Ontario, Canada, operates 24 hours, am
3.357	NOAA WEFAX 24 hours, station NAM, Norfolk, Virginia
3.377	WEFAX, Ankara Turkey at 1600-0030z
3.3775	WEFAX, Guam
3.404	North African region VOLMET weather data for international airline pilots, usb
3.404	West African region VOLMET weather data for international airline pilots, usb
3.407	Hurricane Hunter recon plane to NOAA National Hurricane Center, Miami, Fla., usb
3.413	European region VOLMET weather data for international airline pilots, usb
3.413	VOLMET, Shannon Aeradio, Shannon, Ireland, usb
3.43685	WEFAX, Northwood, Great Britain
3.458	Bangkok, Bombay, Calcutta and Karachi radios VOLMET weather, usb
3.458	Southeast Asian region VOLMET weather data for international airline pilots, usb
3.485	Atlantic Ocean region VOLMET weather data for international airline pilots, usb
3.485	Gander Aeradio VOLMET weather data for international airline pilots, usb
3.485	New York Radio VOLMET weather data for international airline pilots, usb
3.500-4.000	amateur radio 80 meter/75 meter sw radio band in the electromagnetic spectrum
3.520	WEFAX, Belgrade Yugoslavia at 1700-0700z
3.5815	bulletins on the half hour in major comms emergencies, ARRL hq stn W1AW, cw
3.625	bulletins 15 min past the hour in major comm emergencies, ARRL stn W1AW, rtty
3.650	WEFAX, Madrid at 0400-1700z
3.800-3.999	amateur radio, various local nets carrying hurricane traffic as needed, lsb
3.810	standard time, HD210a, Guayaquil, Ecuador, utc operating hours: 0500-1700
3.815	amateur radio, Caribbean weather net, lsb
3.815	amateur radio, National Hurricane Watch, alternate to primary 14.325, lsb
3.855	amateur radio, Great Circle Shortwave Society Net, nationwide, Monday 0100z, lsb

WEATHER FREQUENCIES 91

3.855	WEFAX, Hamburg, Germany at 0900-1000z
3.900	amateur radio, National Hurricane Watch alternate to primary 14.325, lsb
3.900-4.000	75-meter worldwide shortwave broadcasting band, am
3.930	amateur radio, Puerto Rico weather net, lsb
3.9675	amateur radio, U.S. West Coast maritime net, lsb
3.968	amateur radio, Waterways Net, U.S. East Coast, lsb
3.990	bulletins on the hour in major comms emergencies, ARRL hq stn W1AW, lsb
4.0375	WEFAX, Norrkoping, Sweden
4.040	joint military emergency disaster net, usb
4.0466	U.S. Coast Guard, Pacific area, emergency command network, usb
4.0475	WEFAX, Paris
4.048	U.S. Coast Guard, Atlantic area, emergency command network, rtty, cw and ssb usb
4.055	Dept. of Transportation emergency net, nationwide, with FAA fixed net, usb and data
4.125	international voice distress, safety and calling backup, usb
4.2025–4.207	wx rtty to Murmansk by Russian fish fleet off US/Canada Atlantic coast Feb-May
4.223	WEFAX, Delaware, WLO
4.2325	standard time, VPS8, Kowloon, Hong Kong, operates every odd hour 1100-2100 utc
4.244	standard time, PPR, Rio de Janeiro, Brazil, 5-min transm'n at 0125, 1425, 2125 utc
4.2475	standard time, PPR, Rio de Janeiro, Brazil, 5-min transm'n at 0125, 1425, 2125 utc
4.24785	WEFAX, Northwood, Great Britain
4.268	WEFAX, Esquimalt, British Columbia, Canada
4.271	WEFAX, Halifax, Nova Scotia
4.286	standard time, VWC, Calcutta, India, 5-minute transmission at 1625 utc
4.291	standard time, ZSC, Capetown, South Africa, 5-min transmission at 0755, 1655 utc
4.296	WEFAX, Kodiak, Alaska
4.2961	marine weather charts, WEFAX
4.298	standard time, CBV, Valparaiso, Chile, 1155,1555,1955,0055, hr early Oct15-Mar15
4.322	WEFAX, Monsanto at 0635-1700z
4.335	U.S. Coast Guard emergency command network, rtty and cw
4.346	WEFAX, San Francisco at 0100-1500z
4.363	marine weather, 1300z, AT&T high-seas radiotelephone stn WOM, Florida, usb
4.363	marine weather, 2300z, AT&T high-seas radiotelephone stn WOM, Florida, usb
4.376	Canada Coast Guard, Resolute, North West Territories, wx forecast, usb
4.376	Canada, Resolute, NWT weather forecast, usb
4.387	marine weather, 1200z, AT&T high-seas radiotelephone stn WOO, New Jersey, usb
4.387	marine weather, 2200z, AT&T high-seas radiotelephone stn WOO, New Jersey, usb
4.402	marine weather, 0000z, AT&T high-seas radiotelephone stn KMI, California, usb
4.402	marine weather, 1200z, AT&T high-seas radiotelephone stn KMI, California, usb
4.419	weather broadcasts by maritime public coast stn WAK, New Orleans, Louisiana, usb
4.4287	U.S. Coast Guard tropical storm bulletins
4.4287	U.S. Coast Guard, Portsmouth, Va., weather broadcast, usb
4.4287	U.S. Coast Guard, Portsmouth, Virginia, weather broadcast, usb
4.4287	U.S. Coast Guard, tropical storm bulletin, usb
4.4287	U.S. Coast Guard, tropical storm bulletins, usb
4.4318	NOAA to research ships worldwide, ships transmit 4.1374, usb
4.4318	NOAA to research vessels, worldwide, ship transmits 4.1374, usb
4.445	standard time, NPO, Subic Bay, Phillippines, 5-min at 0555, 1155, 1755, 2355 utc
4.4675	Civil Air Patrol communications, hurricane net, Gulf Coast, usb
4.5167	WEFAX, Khabarovsk
4.582	Civil Air Patrol national emergency and calling frequency, usb
4.610	WEFAX, Bracknell, Great Britain
4.6375	hurricane preparations, oil rig evacuation, offshore Gulf of Mexico, usb
4.701	Hurricane Hunter aircraft-to-aircraft, backup frequency, ssb
4.704	WEFAX, station AOK, Rota, Spain
4.722	RAF, West Drayton, England, VOLMET weather data for int'l airline pilots, usb

4.730	RAF Strike Command, Great Britain, weather broadcast, usb
4.730	RAF strike command, Great Britain, weather broadcast, usb
4.750-5.060	60-meter worldwide shortwave broadcasting band, unofficially 4.600-5.100, am
4.768	WEFAX, Rome
4.782	WEFAX, Bracknell, Great Britain
4.793	WEFAX, Washington, DC
4.8135	joint military emergency disaster net, usb
4.855	WEFAX, Hawaii
4.902	Fed. Highway Admin. net, nationwide Fed. Hiway Emerg. Sys., ch. F16, usb & data
4.975	WEFAX, Guam
4.996	standard time, RWM, Moscow, Russia, operates 24 hours
5.000	standard time, ATA, New Delhi, India, operates 1230-0330 utc
5.000	standard time, BPM, Xian, China, utc operating hours: 1400-2400
5.000	standard time, BSF, Chung-Li, Taiwan, operates 24 hours
5.000	standard time, HD210a, Guayaquil, Ecuador, utc operating hours: 1700-1800
5.000	standard time, HLA, Taejon, Korea, operates 24 hours
5.000	standard time, IAM, Rome, Italy, utc hrs: 0730-0830, 1030-1100, hr earlier summer
5.000	standard time, IBF,Turin,Italy, 15-min at :45 past 6-8z,10-17z, 1 hr earlier summer
5.000	standard time, JJY, Tokyo, Japan, standard time, operates 24 hours, am
5.000	standard time, LOL, Buenos Aires, Argentina, hrs: 11-12,14-15,17-18,20-21,23-24z
5.000	standard time, RCH, Tashkent, Uzbekistan, operates 1400-0400 utc
5.000	standard time, VNG, Canberra, Australia, operates 24 hours, am
5.000	standard time, WWV, Fort Collins, Colo., wx, navigation, solar rpts, 24 hrs, am
5.000	standard time, WWVH, Kauai, Hawaii, weather, navigation, solar rpts, 24 hrs, am
5.000	standard time, YVTO, Caracas, Venezuela, operates 24 hours
5.004	standard time, RID, Irkutsk, Russia, operates 24 hours
5.008	Dept. of Transportation emergency net, nationwide, channel F2, usb and data
5.093	WEFAX, Sofia, Bulgaria
5.206	WEFAX, Athens, Greece at 2000-0800z
5.211	Federal Emergency Management Agency nationwide, usb
5.255	Fed. Highway Admin. net, nationwide Fed. Hiway Emerg. Sys., ch. F2, usb and data
5.320	U.S. Coast Guard iceberg patrol warnings, station NIK, Boston, 0018z, 1218z, usb
5.320	U.S. Coast Guard iceberg patrol warnings, station NIK, Boston, 0050z, 1250z, cw
5.320	U.S. Coast Guard, Florida, during hurricanes, usb
5.355	WEFAX, Moscow
5.4225	U.S. 5th District fixed emergency network, usb
5.424	Fed. Highway Admin. net, nationwide Fed. Hiway Emerg. Sys., ch. F21, usb & data
5.430	standard time, BPM, Xian, China, utc operating hours: every 2 hours 1000-1800
5.499	North African region VOLMET weather data for international airline pilots, usb
5.499	West African region VOLMET weather data for international airline pilots, usb
5.547	aeronautical weather
5.562	Hurricane Hunter recon plane to NOAA National Hurricane Center, Miami, Fla., usb
5.589	Middle East region VOLMET weather data for international airline pilots, usb
5.592	Atlantic Ocean region VOLMET weather data for international airline pilots, usb
5.592	Gander Aeradio VOLMET weather data for international airline pilots, usb
5.592	New York Radio VOLMET weather data for international airline pilots, usb
5.592	North Atlantic region VOLMET weather data for international airline pilots, usb
5.592	Shannon Aeradio VOLMET weather data for international airline pilots, usb
5.601	Indian Ocean region VOLMET weather data for international airline pilots, usb
5.640	European region VOLMET weather data for international airline pilots, usb
5.640	Shannon Aeradio VOLMET weather data for international airline pilots, usb
5.673	Southeast Asian region VOLMET weather data for international airline pilots, usb
5.680	international search-and-rescue on-scene, usb
5.690	Lahr, FRG, VOLMET, Canada MACS, usb
5.692	U.S. Coast Guard iceberg patrol helicopter-to-ground & aircraft hf working freq., usb

WEATHER FREQUENCIES

5.696	U.S. Coast Guard iceberg patrol airplane-to-ground and aircraft hf working freq., usb
5.800	WEFAX, Belgrade, Yugoslavia
5.850	WEFAX, Copenhagen, Denmark at 0030-1005z
5.950-6.200	49-meter worldwide shortwave broadcasting band, unofficially 5.850-6.250, am
6.185	WEFAX, Madrid at 0400-1700z
6.215	international voice distress, safety and calling backup, usb
6.2186	NOAA, Woods Hole, Maine, ships callup 0930,1530 est, usb
6.300–6.3115	wx rtty to Murmansk by Russian fish fleet off US/Canada Atlantic coast Feb-May
6.330	WEFAX, station CFH, Halifax, Nova Scotia
6.381	U.S. Coast Guard emergency command network, rtty
6.4365	WEFAX, Northwood, Great Britain
6.5064	U.S. Coast Guard tropical storm bulletin, usb
6.5095	NOAA to research ships worldwide, ships transmit 6.2031, usb
6.5219	U.S. Coast Guard hurricane warnings, weather broadcasts nationwide, usb
6.538	North African region VOLMET weather data for international airline pilots, usb
6.538	West African region VOLMET weather data for international airline pilots, usb
6.577	U.S. Air Force hurricane reconnaissance, usb
6.580	European region VOLMET weather data for international airline pilots, usb
6.604	Atlantic Ocean region VOLMET weather data for international airline pilots, usb
6.604	Gander Aeradio VOLMET weather data for international airline pilots, usb
6.604	New York Radio VOLMET weather data for international airline pilots, usb
6.673	Hurricane Hunter recon plane to NOAA National Hurricane Center, Miami, Fla., usb
6.676	Bangkok, Bombay, Calcutta, Karachi, Singapore radios VOLMET weather, usb
6.676	Southeast Asian region VOLMET weather data for international airline pilots, usb
6.679	Pacific Ocean region VOLMET weather data for international airline pilots, usb
6.693	Edmonton, St. Johns, Trenton, Canada MACS VOLMET, usb
6.705	Edmonton, St. Johns, Trenton, military aviation weather
6.746	Halifax, Canada MACS VOLMET, usb
6.750	U.S. Air Force hurricane reconnaissance, patches, usb
6.753	Edmonton, Lahr, St. Johns, Trenton, military av wx, Canada MACS VOLMET, usb
6.754	U.S. Air Force hurricane reconnaissance, patches, usb
6.790	WEFAX, Ankara, Turkey at 0500-1400z
6.840	standard time, EBC, San Fernando, Spain, operates 1029-1055 utc
6.850	WEFAX, station WLO, Mobile, Alabama, at 0250-2030z
6.872	news photos facsimile, station LRB79, Buenos Aires
6.901	WEFAX, Norrkoping, Sweden
6.944	WEFAX, station CKN, Vancouver, British Columbia
6.968	WEFAX, Esquimalt, British Columbia, Canada
6.9775	NOAA National Weather Service, Caribbean net channel 5, usb
7.000-7.300	amateur radio, 40-meter sw radio-frequency band in the electromagnetic spectrum
7.0475	amateur radio, bulletins on half hour in major emergencies, ARRL stn W1AW, cw
7.095	amateur radio, bulletins 15 min past hour in major emerg., ARRL stn W1AW, rtty
7.100-7.300	41-meter worldwide shortwave broadcasting band, unofficially 7.100-7.600, am
7.165	amateur radio, National Hurricane Watch Net, alternate to primary 14.325, lsb
7.230	amateur radio, Caribbean maritime mobile net, lsb
7.233	amateur radio, recreational vehicle service net, lsb
7.2385	amateur radio, Baja California maritime net, U.S. West Coast, lsb
7.240	amateur radio, ANARC SWL Net, eastern No. America, Sun 1000 eastern time, lsb
7.268	amateur radio, National Hurricane Watch Net, alternate to primary 14.325, lsb
7.268	amateur radio, Waterways Net, U.S. East Coast, lsb
7.290	amateur radio, bulletins on hour in major emergencies, ARRL hq stn W1AW, lsb
7.298	amateur radio, maritime net, U.S. West Coast, lsb
7.335	standard time, CHU, Ottawa, Ontario, Canada, operates 24 hours, am
7.3735	U.S. Dept. of Transportation emergency net, nationwide, channel F3, usb and data
7.417	WEFAX, Rota, Spain

Frequency	Description
7.419	Federal Highway Admin. net, nationwide, Fed. Hiwy Emerg. Sys., ch. F3, usb,data
7.475	SHARES, U.S. federal govt. emergency net, nationwide, FAA, USDoT, usb and data
7.475	WEFAX, Khabarovsk
7.507	U.S. Coast Guard and U.S. Navy hurricane warnings, USCG frequency papa, usb
7.530	WEFAX, station NMF, Boston, Massachusetts
7.5335	U.S. Federal Emergency Management Agency, VIP relocation centr, Wash., DC, usb
7.582	U.S. Dept. of Transportation emergency net, nationwide, channel F4, usb and data
7.5875	WEFAX, Dakar at 2000-0830z
7.600	standard time, HD210a, Guayaquil, Ecuador, utc operating hours: 1800-0500
7.645	WEFAX, Guam
7.710	WEFAX, Frobisher Bay, CA at 1000-2000z, Jul 1-Oct 15
7.7265	Fed. Highway Admin. net, nationwide, Fed. Hiway Emerg. Sys., ch. F26, usb& data
7.743	Fed. Highway Admin. net, nationwide, Fed. Hiway Emerg. Sys., ch. F28, usb& data
7.750	WEFAX, Moscow
7.770	WEFAX, Hawaii
7.7735	U.S. Coast Guard, Miami, hurricane preparations, usb
7.821	Fed. Highway Admin. net, nationwide, Fed. Hiway Emerg. Sys., ch. F29, usb& data
7.880	WEFAX, Hamburg, Germany at 0900-1000z
8.000	standard time, JJY, Tokyo, Japan, standard time, operates 24 hours, am
8.018	WEFAX, Helsinki, Finland at 0740z
8.040	WEFAX, Bracknell, Great Britain
8.0515	marine wx data, even hrs :20, AT&T high-seas radiotelephone WOO, New Jersey
8.075	WEFAX, Norrkoping, Sweden
8.080	NOAA 24-hour WEFAX; station NAM, Norfolk, Virginia
8.087	marine weather data, odd hrs :20, AT&T high-seas radiotelephone KMI, California
8.090	USCG iceberg patrol warning, Norfolk, 0800-0900, 1500-1700, 2100-2200z, usb
8.100	WEFAX, Athens, Greece
8.125	SHARES, U.S. federal govt. emergency net, nationwide, FAA, USDoT, usb and data
8.146	WEFAX, Rome
8.185	WEFAX, Paris
8.260	WEFAX, Monsanto at 0635-1700z
8.291	international voice distress, safety and calling backup, usb
8.2911	NOAA, Woods Hole, Maine, ships callup 0930,1530 est, usb
8.364	international cw and mcw lifeboat, survival craft and search-and-rescue on-scene forces
8.3801	marine weather charts, WEFAX
8.396–8.4145	wx rtty to Murmansk by Russian fish fleet off US/Canada Atlantic coast Feb-May
8.457	WEFAX, Kodiak, Alaska, maritime weather charts
8.461	standard time, ZSC, Capetown, South Africa, 5-min transmission at 0755, 1655 utc
8.473	standard time, 4PB, Colombo, Sri Lanka, operates 0553-0600, 1323-1330 utc
8.492	standard time, PPR, Rio de Janeiro, Brazil, 5-min transm'n at 0125, 1425, 2125 utc
8.494	WEFAX, Alaska
8.4945	WEFAX, Northwood, Great Britain
8.502	U.S. Coast Guard iceberg patrol warnings, station NIK, Boston, 0018z, 1218z, usb
8.502	U.S. Coast Guard iceberg patrol warnings, station NIK, Boston, 0050z, 1250z, cw
8.502	U.S. Coast Guard iceberg patrol warnings, station NIK, Boston, 1600z, fax
8.502	WEFAX, Boston, MA at 1600z, ice reports, March-July
8.502	WEFAX, Norfolk VA
8.539	standard time, VPS35, Kowloon, Hong Kong, operates every odd hour
8.542	standard time, PKX, Jakarta, Indonesia, operates 0045-0100 utc
8.634	standard time, PPR, Rio de Janeiro, Brazil, 5-min transm'n at 0125, 1425, 2125 utc
8.646	WEFAX, San Diego, CA
8.648	U.S. Coast Guard emergency command network, rtty and cw
8.650	standard time, OBC3, Callao, Peru, 5-minute transmission at 1555, 2055, 0155 utc
8.677	standard time, CBV,Valparaiso,Chile,5-min :55 past 00,11,15,19,early Oct15-Mar15
8.680	maritime weather charts, WEFAX

8.682	WEFAX, station NMC, San Francisco, California
8.721	standard time, PPE, Rio de Janeiro, Brazil, 5-min at :25 past 00,11,13,18,20,23 utc
8.722	marine weather, 1300 & 2300z, AT&T high-seas radiotelephone stn WOM, Fla., usb
8.749	marine weather, 1200 & 2200z, AT&T high-seas radiotelephone stn WOO, N.J., usb
8.753	NOAA to research vessels worldwide, ship transmits 8229.1, usb
8.7654	U.S. Coast Guard, New Orleans, Louisiana, weather broadcast, usb
8.7654	U.S. Coast Guard, Portsmouth, Virginia, weather broadcast, usb
8.7685	U.S. Coast Guard tropical storm bulletins, usb
8.828	Pacific Ocean region VOLMET weather data for international airline pilots, usb
8.843	aeronautical weather
8.849	Southeast Asian region VOLMET weather data for international airline pilots, usb
8.852	North African region VOLMET weather data for international airline pilots, usb
8.852	West African region VOLMET weather data for international airline pilots, usb
8.870	Atlantic Ocean region VOLMET weather data for international airline pilots, usb
8.870	Gander Aeradio VOLMET weather data for international airline pilots, usb
8.870	New York Radio VOLMET weather data for international airline pilots, usb
8.870	North Atlantic region VOLMET weather data for international airline pilots, usb
8.870	Shannon Aeradio VOLMET weather data for international airline pilots, usb
8.876	Hurricane Hunter recon plane to NOAA National Hurricane Center, Miami, Fla., usb
8.918	aviation weather and air tfc control in hurricane area, usb
8.945	Middle East region VOLMET weather data for international airline pilots, usb
8.957	European region VOLMET weather data for international airline pilots, usb
8.957	Shannon Aeradio VOLMET weather data for international airline pilots, usb
8.980	U.S. Coast Guard aircraft hf working frequency, usb
8.984	USCG iceberg patrol helicopter-to-ground, airplane-to-ground, usb
8.993	U.S. Air Force hurricane reconnaissance, patches, usb
9.0745	U.S. Dept. of Transportation emergency net, nationwide, channel F5, usb and data
9.092	WEFAX, Brentwood NY at 0712-1212z
9.125	U.S. Coast Guard emergency command network, rtty, cw and usb
9.157	WEFAX, station WLO, Mobile, Alabama
9.169	Fed. Highway Admin. net, nationwide, Fed. Hiwy Emerg. Sys., ch. F30, usb& data
9.197	Fed. Highway Admin. national command net, Fed Hwy Emer Sys, ch. F4, usb&data
9.203	WEFAX, Bracknell, Great Britain
9.230	WEFAX, Khabarovsk
9.351	standard time, BPM, Xian, China, utc operating hrs: 0600 and every hour 1100-2300
9.360	WEFAX, Copenhagen at 0005-1850z
9.380	U.S. Coast Guard and U.S. Navy hurricane warnings, USCG frequency papa, usb
9.3895	WEFAX, Brentwood NY at 0712-1212z
9.440	WEFAX, Hawaii
9.500-9.775	31-meter worldwide shortwave broadcasting band, unofficially 9.250-9.995, am
9.875	WEFAX, Rota, Spain
9.890	WEFAX, Halifax, Nova Scotia
9.9825	WEFAX, Honolulu
9.996	standard time, RWM, Moscow, Russia, operates 24 hours
10.000	standard time, ATA, New Delhi, India, operates 24 hours
10.000	standard time, BPM, Xian, China, operates 24 hours
10.000	standard time, JJY, Tokyo, Japan, standard time, operates 24 hours, am
10.000	standard time, LOL, BuenosAires,Argentina, utc hrs: 11-12,14-15,17-18,20-21,23-24
10.000	standard time, RCH, Tashkent, Uzbekistan, operates 0500-1330 utc
10.000	standard time, RTA, Novosibirsk, Russia, 0200-0500, 1400-1730, 1800-0130 utc
10.000	standard time, VNG, Canberra, Australia, operates 2200-0700z, am
10.000	standard time, WWV, Fort Collins, Colo., wx, navigation, solar rpts, 24 hrs, am
10.000	standard time, WWVH, Kauai, Hawaii, weather, navigation, solar rpts, 24 hrs, am
10.004	standard time, RID, Irkutsk, Russia, operates 24 hours
10.015	Hurricane Hunter recon plane to NOAA National Hurricane Center, Miami, Fla., usb

10.0204	standard time, VPS60, Kowloon, Hong Kong, every odd hour 0100-1500 utc
10.051	Atlantic Ocean region VOLMET weather data for international airline pilots, usb
10.051	Gander Aeradio VOLMET weather data for international airline pilots, usb
10.051	New York Radio VOLMET weather data for international airline pilots, usb
10.057	North African region VOLMET weather data for international airline pilots, usb
10.057	West African region VOLMET weather data for international airline pilots, usb
10.087	Indian Ocean region VOLMET weather data for international airline pilots, usb
10.100-10.150	amateur radio, 30-meter sw radio-frequency band in the electromagnetic spectrum
10.123	WEFAX, Cairo
10.136	U.S. Coast Guard, Atlantic area emergency command network, rtty, cw and usb
10.166	U.S. Coast Guard, Pacific area emergency command network, rtty, cw and usb
10.185	WEFAX, Washington, DC
10.194	Federal Emergency Management Agency, nationwide, channel F27, usb
10.195	SHARES, U.S. federal govt. emergency net, nationwide, usb and data
10.250	WEFAX, Madrid at 0400-1700z
10.255	WEFAX, Guam
10.4405	standard time, NPO, Subic Bay, Phillippines, 5-min at 0555, 1155, 1755, 2355 utc
10.493	Federal Emergency Management Agency, SHARES, nationwide, channel F28, usb
10.535	WEFAX, station CFH, Halifax, Nova Scotia
10.588	Federal Emergency Manag. Agency nationwide, ch. F29, with FHWA, USCG, usb
10.677	WEFAX, station LRN2, Buenos Aires
10.863	NOAA 24-hour WEFAX, station NAM, Norfolk Virginia, schedule sent at 2400z
10.891	Fed. Highway Admin. emerg. command net SHARES nationwide, ch. F5, usb/data
10.918	Fed. Highway Admin. net, nationwide, Fed. Hiwy. Emerg. Sys., ch. F34, usb,data
11.028	SHARES, U.S. federal govt. emergency net, nationwide, usb and data
11.028	U.S. Dept. of Transportation emergency net, nationwide, channel F6, usb and data
11.035	WEFAX, Brentwood NY
11.090	WEFAX, Hawaii
11.197	U.S. Coast Guard aircraft hf working frequency, usb
11.200	RAF, West Drayton, England, VOLMET weather data for int'l airline pilots, usb
11.201	U.S. Coast Guard aircraft hf working frequency, usb
11.246	U.S. Air Force hurricane reconnaissance, usb
11.249	Halifax military aviation weather, Canada MACS VOLMET, usb
11.282	aeronautical weather
11.378	European region VOLMET weather data for international airline pilots, usb
11.387	Bangkok , Bombay , Calcutta , Karachi , Singapore radios VOLMET weather, usb
11.387	Southeast Asian region VOLMET weather data for international airline pilots, usb
11.393	Middle East region VOLMET weather data for international airline pilots, usb
11.396	aviation weather and air tfc control in hurricane area, usb
11.398	Hurricane Hunter recon plane to NOAA National Hurricane Center, Miami, Fla., usb
11.434	U.S. Coast Guard emergency command network, rtty, cw and usb
11.440	Shannon Aeradio, Ireland, weather, fsk data rtty, shift 850 baud 50N
11.440	standard time, PLC, Jakarta, Indonesia, operates 0045-0100 utc
11.513	Shannon Aeradio, Ireland, weather, fsk data rtty, shift 850 baud 50N, and cw
11.5136	U.S. Coast Guard emergency command network, usb
11.5187	Fed. Highway Admin. net, nationwide, Fed. Hiway Emerg. Sys., ch. F36, usb, data
11.700-11.975	25-meter worldwide shortwave broadcasting band, unofficially 11.500-12.100, am
12.008	standard time, EBC, San Fernando, Spain, operates 0959-1025 utc
12.135	U.S. Coast Guard hurricane warnings, cw
12.135	USCG iceberg patrol warnings, Norfolk, 0800-0900z, 1500-1700z, 2100-2200z, usb
12.148	U.S. Coast Guard, Atlantic area, emergency command network, rtty, cw and usb
12.158	Fed. Hiwy Admn. command net, nationwide, Fed Hiwy Emerg Sys, ch F6, usb,data
12.173	U.S. Coast Guard, Pacific area, emergency command network, rtty, cw and usb
12.1787	Federal Highway Admin. net, nationwide, Fed Hiwy Emerg Sys, ch. F41, usb& data
12.201	WEFAX, Washington, DC

WEATHER FREQUENCIES 97

12.290	international voice distress, safety and calling backup, usb
12.307	standard time, OBC3, Callao, Peru, 5-minute transmission at 1555, 2055, 0155 utc
12.4292	NOAA, Woods Hole, Maine, vessels callup 0930,1530 est, usb
12.56–12.5765	wx rtty to Murmansk by Russian fish fleet off US/Canada Atlantic coast Feb-May
12.687	standard time, PPR, Rio de Janeiro, Brazil, 5-min transm'n at 0125, 1425, 2125 utc
12.710	WEFAX, Japan
12.724	standard time, ZSC, Capetown, South Africa, 5-minute transm'n at 0755, 1655 utc
12.7281	marine weather charts, WEFAX
12.730	WEFAX, San Francisco
12.738	standard time, PPR, Rio de Janeiro, Brazil, 5-min transm'n at 0125, 1425, 2125 utc
12.745	standard time, VWC, Calcutta, India, 5-minute transmission at 0825 utc
12.750	U.S. Coast Guard iceberg patrol warnings, station NIK, Boston, 0018z, 1218z, usb
12.750	U.S. Coast Guard iceberg patrol warnings, station NIK, Boston, 0050z, 1250z, cw
12.750	U.S. Coast Guard iceberg patrol warnings, station NIK, Boston, 1600z, fax
12.750	WEFAX, Boston
12.804	standard time, NPO, Subic Bay, Phillippines, 5-min at 0555, 1155, 1755, 2355 utc
12.8875	U.S. Coast Guard emergency command network, rtty and cw
13.083	marine wx, 0000z&1200z, AT&T high-seas radiotelephone stn KMI, California, usb
13.092	marine wx, 1300z&2300z, AT&T high-seas radiotelephone stn WOM, Florida, usb
13.110	marine weather, coastal station WLO, Mobile, Alabama, channel 1212
13.1122	U.S. Coast Guard, Honolulu, Hawaii, weather broadcast, usb
13.1132	U.S. Coast Guard tropical storm bulletins, usb
13.1411	NOAA to research vessels worldwide, ships transmit 12.3703, usb
13.231	Lahr, FRG, military aviation weather, VOLMET, Canada MACS, usb
13.260	U.S. Coast Guard and U.S. Navy hurricane warnings
13.261	aeronautical weather
13.261	North African region VOLMET weather data for international airline pilots, usb
13.261	West African region VOLMET weather data for international airline pilots, usb
13.264	European region VOLMET weather data for international airline pilots, usb
13.264	Shannon Aeradio VOLMET weather data for international airline pilots, usb
13.267	Hurricane Hunter recon plane to NOAA National Hurricane Center, Miami, Fla., usb
13.270	Atlantic Ocean region VOLMET weather data for international airline pilots, usb
13.270	Gander Aeradio VOLMET weather data for international airline pilots, usb
13.270	New York Radio VOLMET weather data for international airline pilots, usb
13.270	North Atlantic region VOLMET weather data for international airline pilots, day, usb
13.270	Shannon Aeradio VOLMET weather data for international airline pilots, day, usb
13.276	Atlantic Ocean region VOLMET weather data for international airline pilots, usb
13.276	Gander Aeradio VOLMET weather data for international airline pilots, usb
13.276	New York Radio VOLMET weather data for international airline pilots, usb
13.279	Indian Ocean region VOLMET weather data for international airline pilots, usb
13.282	Pacific Ocean region VOLMET weather data for international airline pilots, usb
13.285	Southeast Asian region VOLMET weather data for international airline pilots, usb
13.300	aeronautical weather
13.345	aeronautical weather
13.352	South American VOLMET weather data for international airline pilots, usb
13.354	Hurricane Hunter recon plane to NOAA National Hurricane Center, Miami, Fla., usb
13.4325	U.S. Dept. of Transportation emergency net, nationwide, channel F7, usb and data
13.510	WEFAX, Halifax, Nova Scotia
13.600-13.800	22-meter worldwide shortwave broadcasting band, am
13.630	SHARES, U.S. federal govt. emergency net, nationwide, FAA, USDoT, usb and data
13.751	news photos facsimile, London
13.862	WEFAX, Hawaii
14.000-14.350	amateur radio, 20-meter sw radio-frequency band in the electromagnetic spectrum
14.0475	amateur radio, bulletins on half hour in major emergencies, ARRL stn W1AW, cw
14.095	amateur radio, bulletins 15 min past hour in major emerg., ARRL stn W1AW, rtty

Frequency	Description
14.265-14.325	amateur radio, nets during emergencies carrying weather, relief and maritime tfc, usb
14.270	amateur radio, Red Cross in hurricanes, usb
14.283	amateur radio, Friendly Caribus Connection weather and disaster net, usb
14.290	amateur radio, bulletins on hour in major emergencies, ARRL hq stn W1AW, usb
14.303	amateur radio, International Assistance and Traffic Net, usb
14.313	amateur radio, maritime miobile nets, 24 hours, usb
14.325	amateur radio, Nat'l Hurricane Watch, Nat'l Hurricane Ctr, Coral Gables, Fla., usb
14.342	amateur radio, U.S. West Coast maritime and weather nets, usb
14.450	Federal Emerg. Management Agency, Denver, Colo., usb & rtty, shift 850 baud75N
14.450	Federal Emerg. Mgnt Agency, San Francisco, Calif., usb & rtty, shift 850 baud75N
14.450	Federal Emergency Management Agency, nationwide, channel F35, usb
14.461	Federal Highway Admin. net, nationwide, Fed Hiway Emerg Sys, ch. F7, usb & data
14.498	Santa Maria, Arizona, weather, fsk data rtty, shift 850 baud 50N
14.670	standard time, CHU, Ottawa, Ontario, Canada, operates 24 hours, am
14.6715	WEFAX, Washington, DC
14.776	Federal Emergency Management Agency, nationwide, channel F36, usb
14.828	WEFAX, station NPM, Pearl Harbor, Hawaii
14.902	Civil Air Patrol, nationwide, with U.S. Air Force, usb
14.902	Federal Emergency Management Agency, Sacramento, Calif., usb
14.902	National Chaplains' net, 2130 utc, usb
14.996	standard time, RWM, Moscow, Russia, operates 24 hours
15.000	standard time, ATA, New Delhi, India, operates 0330-1230 utc
15.000	standard time, BPM, Xian, China, utc operating hours: 0000-1400
15.000	standard time, BSF, Chung-Li, Taiwan, operates 24 hours
15.000	standard time, JJY, Japan, standard time signals, weather, am
15.000	standard time, JJY, Tokyo, Japan, standard time, operates 24 hours, am
15.000	standard time, LOL, Buenos Aires, Argentina, hrs: 11-12,14-15,17-18,20-21,23-24z
15.000	standard time, RTA, Novosibirsk, Russia, operates 0630-0930, 1000-1330 utc
15.000	standard time, VNG, Canberra, Australia, operates 2200-0700z, am
15.000	standard time, WWV, Fort Collins, Colo., wx, navigation, solar rpts, 24 hrs, am
15.000	standard time, WWVH, Kauai, Hawaii, weather, navigation, solar rpts, 24 hrs, am
15.004	standard time, RID, Irkutsk, Russia, operates 24 hours
15.035	Edmonton, St. Johns, Trenton, military aviation wx, Canada MACS VOLMET, usb
15.100-15.450	19-meter worldwide shortwave broadcasting band, unofficially 15.005-15.700, am
15.910	Fed. Highway Admin. net, nationwide, Fed. Hiway Emerg. Sys., ch. F45, usb& data
16.1331	marine weather charts, Pacific and Hawaii, WEFAX
16.180	USCG iceberg patrol warning, Norfolk, 0800-0900, 1500-1700, 2100-2200z, usb
16.201	Federal Emergency Management Agency, Mt. Weather, Virginia, usb
16.2115	Federal Highway Admin. net, nationwide, Fed Hiway Emerg Sys, ch. F8, usb & data
16.348	SHARES, U.S. federal govt. emergency net, nationwide, FAA, USDoT, usb and data
16.410	NOAA WEFAX 0900-2100z, station NAM, Norfolk, Virginia
16.420	international voice distress, safety and calling backup, usb
16.5871	NOAA, Woods Hole, Maine, vessels callup 0930,1530 est, usb
16.785-16.804	wx rtty to Murmansk by Russian fish fleet off US/Canada Atlantic coast Feb-May
17.018	standard time, ZSC, Capetown, South Africa, 5-minute transm'n at 0755, 1655 utc
17.096	standard time, VPS80, Kowloon, Hong Kong, every odd hour 2100-1300 utc
17.1512	WEFAX, San Francisco
17.1944	standard time, PPR, Rio de Janeiro, Brazil, 5-min transm'n at 0125, 1425, 2125 utc
17.242	marine wx, 1300z & 2300z, AT&T high-seas radiotelephone stn WOM, Florida, usb
17.267	NOAA to research vessels worldwide, ships transmit 16.4941, usb
17.3073	U.S. Coast Guard storm warnings nationwide, tropical storm bulletins, usb
17.4105	WEFAX, San Diego
17.421	U.S. Dept. of Transportation emergency net, nationwide, channel F8, usb and data
17.4475	WEFAX, Mobile, Alabama
17.525	Federal Highway Admin net, nationwide, Fed Hiway Emerg Sys, ch. F49, usb& data

17.585	WEFAX, station AOK, Rota, Spain
17.670	news photos facsimile, station LQZ67, Buenos Aires
17.700-17.900	16-meter worldwide shortwave broadcasting band, unofficially 17.500-17.900, am
17.901	Hurricane Hunter recon plane to NOAA National Hurricane Center, Miami, Fla., usb
18.023	Federal Emergency Management Agency, Olney, Maryland, usb
18.068-18.168	amateur radio, 17-meter sw radio-frequency band in the electromagnetic spectrum
18.431	news photos facsimile, LRO83, Buenos Aires
18.893–18.898	wx rtty to Murmansk by Russian fish fleet off US/Canada Atlantic coast Feb-May
19.223	Federal Highway Admin. net, nationwide, Fed Hiway Emerg Sys, ch. F9, usb & data
20.000	standard time, WWV, Fort Collins, Colo., wx, navigation, solar rpts, 24 hrs, am
20.015	NOAA WEFAX transmitted 1200-2100z; Norfolk, Virginia
20.192	NOAA, Jupiter, Florida, control South Atlantic space launch weather net, ssb
20.225	USCG iceberg patrol warning, Norfolk, 0800-0900, 1500-1700, 2100-2200z, usb
20.246	WEFAX, Cairo
20.500	standard time, UQC3, Khabarovsk, Russia, 0036-0117, 0636-0717, 1736-1817 utc
20.500	standard time, UTR3, Gorki, Russia, operates 0536-0617, 1336-1417, 1836-1917 utc
21.000-21.450	amateur radio, 15-meter sw radio-frequency band in the electromagnetic spectrum
21.0775	amateur radio, bulletins on half hour in major emergencies, ARRL stn W1AW, cw
21.095	amateur radio, bulletins 15 min past hour in major emerg., ARRL stn W1AW, rtty
21.272	Federal Emergency Management Agency, nationwide, usb
21.390	amateur radio, bulletins on hour in major emergencies, ARRL hq stn W1AW, usb
21.402	amateur radio, maritime mobile nets, usb
21.450-21.750	13-meter worldwide shortwave broadcasting band, unofficially 21.450-21.850, am
21.937	Hurricane Hunter recon plane to NOAA National Hurricane Center, Miami, Fla., usb
22.124	NOAA, Woods Hole, Maine, vessels callup 0930,1530 est, usb
22.352	standard time, PPR, Rio de Janeiro, Brazil, 5-min transm'n at 0125, 1425, 2125 utc
22.352–22.374	wx rtty to Murmansk by Russian fish fleet off US/Canada Atlantic coast Feb-May
22.420	standard time, PPR, Rio de Janeiro, Brazil, 5-min transm'n at 0125, 1425, 2125 utc
22.455	standard time, ZSC, Capetown, South Africa, 5-minute transm'n at 0755, 1655 utc
22.536	standard time, VPS22, Kowloon, Hong Kong, every odd hour 0100-0900 utc
22.542	news photos facsimile, station JJC, Tokyo, Japan
22.6952	NOAA to research vessels worldwide, ships transmit 22.0992, usb
22.738	marine wx, 1300z & 2300z, AT&T high-seas radiotelephone stn WOM, Florida, usb
23.000	standard time, UQC3, Khabarovsk, Russia, 0036-0117, 0636-0717, 1736-1817 utc
23.000	standard time, UTR3, Gorki, Russia, operates 0536-0617, 1336-1417, 1836-1917 utc
23.3294	WEFAX, U.S.
24.890-24.990	amateur radio, 12-meter sw radio-frequency band in the electromagnetic spectrum
25.000	standard time, UQC3, Khabarovsk, Russia, 0036-0117, 0636-0717, 1736-1817 utc
25.000	standard time, UTR3, Gorki, Russia, operates 0536-0617, 1336-1417, 1836-1917 utc
25.100	standard time, UQC3, Khabarovsk, Russia, 0036-0117, 0636-0717, 1736-1817 utc
25.100	standard time, UTR3, Gorki, Russia, operates 0536-0617, 1336-1417, 1836-1917 utc
25.193–25.208	wx rtty to Murmansk by Russian fish fleet off US/Canada Atlantic coast Feb-May
25.378	U.S. Coast Guard emergency command network, rtty and cw
25.500	standard time, UQC3, Khabarovsk, Russia, 0036-0117, 0636-0717, 1736-1817 utc
25.500	standard time, UTR3, Gorki, Russia, operates 0536-0617, 1336-1417, 1836-1917 utc
25.600-26.100	11-meter worldwide shortwave broadcasting band, am
26.620	Civil Air Patrol communications, am
26.965	Citizens' Band (CB) radio channel 1, am and ssb
26.975	Citizens' Band (CB) radio channel 2, am and ssb
26.985	Citizens' Band (CB) radio channel 3, am and ssb
27.005	Citizens' Band (CB) radio channel 4, am and ssb
27.015	Citizens' Band (CB) radio channel 5, am and ssb
27.025	Citizens' Band (CB) radio channel 6, am and ssb
27.035	Citizens' Band (CB) radio channel 7, am and ssb
27.055	Citizens' Band (CB) radio channel 8, agriculture, am and ssb

27.065	Citizens' Band (CB) radio emergency channel 9, REACT teams, am and ssb
27.075	Citizens' Band (CB) radio channel 10, am and ssb
27.085	Citizens' Band (CB) radio channel 11, am and ssb
27.105	Citizens' Band (CB) radio channel 12, am and ssb
27.115	Citizens' Band (CB) radio channel 13, boats and RVs, am and ssb
27.125	Citizens' Band (CB) radio channel 14, am and ssb
27.135	Citizens' Band (CB) radio channel 15, am and ssb
27.155	Citizens' Band (CB) radio channel 16, am and ssb
27.165	Citizens' Band (CB) radio channel 17, am and ssb
27.175	Citizens' Band (CB) radio channel 18, am and ssb
27.185	Citizens' Band (CB) radio emergency channel 19, truckers, am and ssb
27.205	Citizens' Band (CB) radio channel 20, am and ssb
27.215	Citizens' Band (CB) radio channel 21, am and ssb
27.225	Citizens' Band (CB) radio channel 22, am and ssb
27.235	Citizens' Band (CB) radio channel 24, am and ssb
27.245	Citizens' Band (CB) radio channel 25, am and ssb
27.255	Citizens' Band (CB) radio channel 23, am and ssb
27.265	Citizens' Band (CB) radio channel 26, am and ssb
27.275	Citizens' Band (CB) radio channel 27, am and ssb
27.285	Citizens' Band (CB) radio channel 28, am and ssb
27.295	Citizens' Band (CB) radio channel 29, am and ssb
27.305	Citizens' Band (CB) radio channel 30, am and ssb
27.315	Citizens' Band (CB) radio channel 31, am and ssb
27.325	Citizens' Band (CB) radio channel 32, am and ssb
27.335	Citizens' Band (CB) radio channel 33, am and ssb
27.345	Citizens' Band (CB) radio channel 34, am and ssb
27.355	Citizens' Band (CB) radio channel 35, am and ssb
27.365	Citizens' Band (CB) radio channel 36, am and ssb
27.375	Citizens' Band (CB) radio channel 37, am and ssb
27.385	Citizens' Band (CB) radio channel 38, am and ssb
27.395	Citizens' Band (CB) radio channel 39, am and ssb
27.405	Citizens' Band (CB) radio channel 40, am and ssb
28.000-29.700	amateur radio, 10-meter sw radio-frequency band in the electromagnetic spectrum
28.0775	amateur radio, bulletins on half hour in major emergencies, ARRL stn W1AW, cw
28.095	amateur radio, bulletins 15 min past hour in major emerg., ARRL stn W1AW, rtty
28.590	amateur radio, bulletins on hour in major emergencies, ARRL hq stn W1AW, usb
30.660-31.140	local municipal bus systems
31.480-33.380	hurricane preparation, offshore Gulf of Mexico, nbfm
34.900	National Guard and U.S. Army communications in local emergencies
39.980	air pollution warnings, 1900z, So. Coast Air Quality Mngmnt Dist., Calif., nbfm
40.500	U.S. Army distress net, U.S. Dept. of Defense joint operations common frequency
41.950	private maritime coastal stations and military in search and rescue, nbfm
42.080-42.340	hurricane preparation, South Carolina police, nbfm
43.700-43.840	local passenger bus systems, nbfm
44.460-44.600	local municipal bus systems, nbfm
47.420	Red Cross in disasters, nbfm
47.460	Red Cross in disasters, nbfm
47.500	Red Cross in disasters, nbfm
47.660	Red Cross in disasters, nbfm
50.000	NOAA wind-profiler radar to detect very high wind shear, planned for future
50.000-54.000	amateur radio, 6-meter vhf radio-frequency band in the electromagnetic spectrum
50.080	bulletins on the half hour in major comms emergencies, ARRL hq stn W1AW, cw
50.190	bulletins on the hour in major comms emergencies, ARRL hq stn W1AW, ssb
59.750	TV audio, U.S. television broadcast channel 2
65.750	TV audio, U.S. television broadcast channel 3

Frequency	Description
71.750	TV audio, U.S. television broadcast channel 4
81.750	TV audio, U.S. television broadcast channel 5
87.750	TV audio, U.S. television broadcast channel 6
88.00-108.000	broadcast stations, some carrying weather and marine forecasts and warnings, wbfm
108.00-136.000	Automated Terminal Information Service (ATIS), airport weather broadcasts
108.00-136.000	Automated Weather Observing System (AWOS), airport weather broadcasts
108.400	Philipps Army Air Field, Md., wx, Automated Terminal Information Service (ATIS)
118.375	Portsmouth, Va., airport weather, Automated Weather Observing System (AWOS)
118.425	Charlottesville, Va., airport wx, Automated Weather Observing System (AWOS)
118.450	Martinsville, Va., airport weather, Automated Weather Observing System (AWOS)
118.600	Wise, Va., airport weather, Automated Weather Observing System (AWOS)
118.800	Hot Springs, Va., airport weather, Automated Weather Observing System (AWOS)
119.800	Lynchburg, Va., airport weather, Automated Weather Observing System (AWOS)
121.500	civilian aircraft and ship emergencies, usb, Class A and B EPIRB and ELT xmtrs
121.600	U.S. and Canada search-and-rescue on-scene, voice
122.000	U.S. aviation weather nationwide
122.900	search and rescue operations, local and regional
123.050	Hurricane Hunter aircraft-to-aircraft, primary frequency, am
123.100	international search-and-rescue on-scene, primary frequency, voice
123.300	balloons
123.600	local airport weather advisories
123.950	New Castle County, Delaware, wx, Automated Terminal Information Service (ATIS)
124.850	Winchester, Va., airport weather, Automated Weather Observing System (AWOS)
124.925	Shenandoah Valley, Va., airport wx, Automated Weather Observing System (AWOS)
127.525	Manassas, Va., airport weather, Automated Weather Observing System (AWOS)
128.625	Chesterfield, Va., airport weather, Automated Weather Observing System (AWOS)
128.850-132.00	airline companies two-way communications
133.325	Petersburg, Va., airport weather, Automated Weather Observing System (AWOS)
136.000-138.00	WEFAX from low-orbit weather satellites
136.110	WEFAX from MOS-1B, Earth-observation satellite, Japan
136.370	ATS-3 sat, hurricane hunter aircraft to Nat'l Hurricane Center, Coral Gables, voice
136.380	WEFAX from old GOES-1, GOES-2, GOES-3 weather satellites, U.S.
136.770	WEFAX from NOAA-6, NOAA-8, NOAA-9, NOAA-11 weather satellites, U.S.
137.050	WEFAX from Meteosat 1 weather satellite, Europe
137.076	WEFAX from Meteosat 2 weather satellite, Europe
137.080	WEFAX from weather satellite Meteosat, Europe
137.300	WEFAX from Meteor 2-14, 2-17, 3-03 and 3-04 weather satellites, Russia
137.400	WEFAX from Meteor 2-13 and 2-16 weather satellites, Russia
137.400	WEFAX from Okean 2 satellite, Russia
137.500	WEFAX from NOAA-10, NOAA-12 weather satellites, U.S.
137.620	WEFAX from NOAA-9, NOAA-11 weather satellites, U.S.
137.795	WEFAX from Feng Yun 1B weather satellite, China
137.850	WEFAX from Meteor 2-14, 2-15, 2-19 and 3-01 weather satellites, Russia
138.00-144.00	military operations band, police, security, medical, fire, safety, others
138.450	on-the-scene search-and-rescue, fm
138.780	on-the-scene search-and-rescue, fm
141.080	military local and regional civil emergency operations
141.120	military local and regional civil emergency operations
141.465	military local and regional civil emergency operations
142.230	Federal Emergency Management Agency (FEMA), civil defense (CD), nbfm
142.350	Federal Emergency Management Agency (FEMA), civil defense (CD), nbfm
142.425	Federal Emergency Management Agency (FEMA), civil defense (CD), nbfm
142.440	military local and regional civil emergency operations
142.975	Federal Emergency Management Agency (FEMA), civil defense (CD), nbfm
143.000	Federal Emergency Management Agency (FEMA), civil defense (CD), nbfm

Frequency	Description
143.900	Civil Air Patrol, am and fm repeater inputs
144.00-148.00	amateur radio, two-meter vhf radio-frequency band in the electromagnetic spectrum
144.00-146.00	amateur radio satellite downlinks
145.00-148.00	amateur radio, 2-meter repeaters, local and regional comms, voice and packet, nbfm
145.00-148.00	amateur radio, National Weather Service local and regional SKYWARN nets, nbfm
145.010	amateur radio, Hurricane Hunter Aircraft Packet Project (HHAPP), digipeater, nbfm
145.490	amateur radio, The Shortwave Listener's Net, central New York, Sun 0100z, nbfm
147.105	amateur radio, The Shortwave Listener's Net, central New York, Sun 0100z, nbfm
147.555	amateur radio, bulletins 15 min past hour in major emerg., ARRL stn W1AW, rtty
147.555	amateur radio, bulletins on half hour in major emergencies, ARRL stn W1AW, cw
147.555	amateur radio, bulletins on hour in major emergencies, ARRL hq stn W1AW, fm
148.00-150.00	military operations, including police, security, medical, fire, safety, others
148.150	Civil Air Patrol, am and fm repeater outputs
148.305	U.S. Coast Guard auxiliary, nbfm
148.825	U.S. Coast Guard auxiliary, nbfm
148.925	Civil Air Patrol, am, nbfm
149.990	Military Affiliate Radio System (MARS), wx emergencies, search & rescue, nbfm
150.81-150.965	local automobile clubs and tow trucks, nbfm
150.995-151.13	highway maintenance crews, road departments, snow removal trucks, nbfm
151.14-151.475	environmental response units, park rangers, game wardens, forestry workers, nbfm
151.49-151.595	large construction firms, large farms, nbfm
152.27-152.450	taxi cab dispatchers, nbfm
152.87-153.725	winter-heat fuel oil delivery, construction, manufacturers, mills, movie crews, nbfm
153.74-156.030	public safety, police, fire, ambulance, hospital, vets, school buses, municipal, nbfm
155.160	special emergency search and rescue, nbfm
155.325-155.40	local emergency medical service (ems) communications, nbfm
155.475	nationwide police emergency channel, nbfm
156.025	maritime vhf communications channel 60, nbfm
156.045-156.24	highway maintenance crews, road departments, snow removal trucks, nbfm
156.050	maritime vhf communications channel 1, VTS, nbfm
156.075	maritime vhf communications channel 61, nbfm
156.100	maritime vhf communications channel 2, port operations, nbfm
156.125	maritime vhf communications channel 62, nbfm
156.150	maritime vhf communications channel 3, port operations, nbfm
156.175	maritime vhf communications channel 63, nbfm
156.200	maritime vhf communications channel 4, port operations, nbfm
156.225	maritime vhf communications channel 64, nbfm
156.250	maritime vhf communications channel 5, VTS, nbfm
156.255	oil spill cleanup, nbfm
156.275	maritime vhf communications channel 65, port operations, nbfm
156.300	maritime vhf channel 6, intership safety & merchant ship/USCG search/rescue, nbfm
156.325	maritime vhf communications channel 66, port operations, nbfm
156.350	maritime vhf communications channel 7, commercial, nbfm
156.375	maritime vhf communications channel 67, commercial vessels, nbfm
156.400	maritime vhf communications channel 8, commercial, nbfm
156.425	maritime vhf communications channel 68, pleasure boats & non-commercial, nbfm
156.450	maritime vhf comm channel 9, commercial and non-commercial, nbfm
156.475	maritime vhf comm channel 69, pleasure boats & non-commercial, nbfm
156.500	maritime vhf communications channel 10, commercial, nbfm
156.525	maritime vhf communications channel 70, non-commercial operations, nbfm
156.550	maritime vhf communications channel 11, commercial, nbfm
156.575	maritime vhf communications channel 71, non-commercial operations, nbfm
156.600	maritime vhf communications channel 12, Coast Guard and port operations, nbfm
156.625	maritime vhf communications channel 72, non-commercial, intership, nbfm
156.650	maritime vhf communications channel 13, navigational operations, nbfm

Frequency	Description
156.675	maritime vhf communications channel 73, port operations, nbfm
156.700	maritime vhf communications channel 14, port operations, nbfm
156.725	maritime vhf communications channel 74, port operations, nbfm
156.750	maritime vhf channel 15, distress/safety rcv; EPIRB 15-sec. homing signal, nbfm
156.800	maritime vhf channel 16, distress/safety and calling, and EPIRB alert tones, nbfm
156.850	maritime vhf communications channel 17, state control operations, nbfm
156.875	maritime vhf communications channel 77, port operations, nbfm
156.900	maritime vhf communications channel 18, commercial, nbfm
156.925	maritime vhf comm channel 78, non-commercial operations, working, nbfm
156.950	maritime vhf communications channel 19, commercial, nbfm
156.975	maritime vhf communications channel 79, commercial, nbfm
157.000	maritime vhf communications channel 20, port operations, nbfm
157.025	maritime vhf communications channel 80, commercial, nbfm
157.050	maritime vhf comm channel 21, Coast Guard, U.S. government operations, nbfm
157.075	maritime vhf comm channel 81, U.S. govt. operations only, oil spill cleanups, nbfm
157.100	maritime vhf comm channel 22, Coast Guard operations, comms to boaters, nbfm
157.125	maritime vhf comm channel 82, U.S. government operations only, nbfm
157.150	maritime vhf comm channel 23, Coast Guard, U.S. government operations, nbfm
157.175	maritime vhf comm channel 83, Coast Guard, U.S. government operations, nbfm
157.425	maritime vhf communications channel 88, commercial operations, nbfm
157.47-157.515	tow trucks, nbfm
157.53-157.710	taxi cab drivers, nbfm
158.13-158.265	power companies, water companies, utility companies, nbfm
158.28-158.445	winter-heat fuel oil delivery, construction, manufacturers, mills, nbfm
158.445	oil spill cleanup, nbfm
158.73-159.210	police, municipalities, highway crews, road departments, snow trucks, nbfm
159.22-159.465	environmental response units, park rangers, nbfm
159.480	oil spill cleanup, nbfm
159.495-160.20	trucking companies, armored cars, nbfm
160.21-161.565	railroad wx, railroad police, yards, local & over-the-road RR communications, nbfm
161.580	oil spill cleanup, nbfm
161.64-161.760	local radio and television news crews, traffic helicopters, weatherman remotes, nbfm
161.650	Canadian weather radio broadcast channel W8, nbfm
161.775	Canadian weather radio broadcast channel W9, nbfm
162.075	National Weather Service, nbfm
162.150	National Weather Service, nbfm
162.400	NOAA weather radio broadcast channel W2, nbfm
162.425	NOAA weather radio broadcast channel W4, nbfm
162.450	NOAA weather radio broadcast channel W5, nbfm
162.475	NOAA weather radio broadcast channel W3, nbfm
162.500	NOAA weather radio broadcast channel W6, nbfm
162.525	NOAA weather radio broadcast channel W7, nbfm
162.550	NOAA weather radio broadcast channel W1, nbfm
163.225	National Weather Service, nbfm
163.275	National Weather Service and National Storms Lab, Norman, Oklahoma, nbfm
163.300	National Weather Service, nbfm
163.325	National Weather Service, nbfm
163.350	National Weather Service, nbfm
163.41-163.437	U.S. Army Corps of Engineers, nbfm
163.4125	U.S. Army Corps of Engineers, nbfm
163.4325	U.S. Army Corps of Engineers, nbfm
163.4875	National Guard communications in local emergencies
163.5125	military local and regional disaster preparedness
164.8625	Federal Emergency Management Agency (FEMA), civil defense (CD), nbfm
165.2625	U.S. Coast Guard internal communications, nbfm

Frequency	Description
165.3125	U.S. Coast Guard relay of marine channel 16, distress/safety and calling, nbfm
165.4125	U.S. Environmental Protection Agency helicopters
165.5375	National Weather Service, nbfm
165.5875	National Weather Service, nbfm
165.750	U.S. National Transportation Safety Board air-crash investigators, nbfm
166.025	National Weather Service, nbfm
169.025	National Weather Service, nbfm
169.075	National Weather Service, nbfm
169.25-169.375	U.S. Federal Aviation Administration units, nbfm
169.425	land-mobile meteorological use by IW, IP, IF, IS
169.450	land-mobile meteorological use by IW, IP, IF, IS, IB, LR
169.475	land-mobile meteorological use by IW, IP, IF, IS, IB, LR
169.500	land-mobile meteorological use by IW, IP, IF, IS, IB, LR
169.525	land-mobile meteorological use by IW, IP, IF, IS, IB, LR
170.225	land-mobile meteorological use by IW, IP, IF, IS, IB, LR
170.250	land-mobile meteorological use by IW, IP, IF, IS, IB, LR
170.275	land-mobile meteorological use by IW, IP, IF, IS, IB, LR
170.300	land-mobile meteorological use by IW, IP, IF, IS, IB, LR
170.325	land-mobile meteorological use by IW, IP, IF, IS, IB, LR
171.025	land-mobile meteorological use by IW, IP, IF, IS, IB, LR
171.050	land-mobile meteorological use by IW, IP, IF, IS, IB, LR
171.075	land-mobile meteorological use by IW, IP, IF, IS, IB, LR
171.100	land-mobile meteorological use by IW, IP, IF, IS, IB, LR
171.125	land-mobile meteorological use by IW, IP, IF, IS, IB, LR
171.3375	U.S. Coast Guard relay of marine channel 16, distress/safety and calling, nbfm
171.825	land-mobile meteorological use by IW, IP, IF, IS, IB, LR
171.850	land-mobile meteorological use by IW, IP, IF, IS, IB, LR
171.875	land-mobile meteorological use by IW, IP, IF, IS, IB, LR
171.900	land-mobile meteorological use by IW, IP, IF, IS, IB, LR
171.925	land-mobile meteorological use by IW, IP, IF, IS, IB, LR
172.025	National Weather Service, nbfm
173.22-173.375	local newspaper reporters, nbfm
173.225	local newspaper communications
173.375	local newspaper communications
179.750	TV audio, U.S. television broadcast channel 7
185.750	TV audio, U.S. television broadcast channel 8
191.750	TV audio, U.S. television broadcast channel 9
197.750	TV audio, U.S. television broadcast channel 10
203.750	TV audio, U.S. television broadcast channel 11
209.750	TV audio, U.S. television broadcast channel 12
215.750	TV audio, U.S. television broadcast channel 13
222.00-225.00	amateur radio, 1.25-meter vhf radio-frequency band in the electromagnetic spectrum
222.00-225.00	amateur radio, National Weather Service local and regional SKYWARN nets, nbfm
222.00-225.00	amateur radio, repeaters, local and regional communications, voice and packet, nbfm
237.900	U.S. Coast Guard air and sea rescue
239.800	Federal Aviation Administration weather
243.000	military aircraft & ship emergency, international survival craft, EPIRB & ELT xmtrs
259.000	continental U.S. air rescue operations, uhf am
272.700	military flights weather
282.800	joint/combined search-and-rescue direction-finding and on-scene primary frequency
287.800	U.S. Coast Guard air and sea rescue
304.800	Hurricane Hunter aircraft-to-aircraft, secondary frequency, worldwide, am
342.500	Federal Aviation Administration (FAA) weather; military flights weather
344.600	Federal Aviation Administration (FAA) weather
350.600	NOAA Cape Radio, Point Malabar, Florida, NASA-related weather, am

Frequency	Description
375.200	PMSV metropolitan weather reports, Edwards AFB, fm
381.000	continental U.S. air rescue operations
381.800	U.S. Coast Guard aircraft working frequency
383.900	U.S. Coast Guard air and sea rescue
400.00-400.15	satellites relaying standard time signals
400.15-412.00	meteorological and research satellites, WEFAX, telemetry, data
403.00-405.00	radiosonde, radio-equipped meteorological research balloons, weather balloons
404.000	NOAA wind-profiler radar, detect wind shear, 31 locations in CT, AR, CO, TX, NE
405.000	aircraft-to-GOES satellite reconnaissance data uplink from NOAA P-3 aircraft
406.000	SARSAT/COSPAS satellite uplink, including GOES, in aircraft and ship distress
409.750	U.S. National Storms Laboratory, Norman, Oklahoma, nbfm
410.075	NOAA-to-local station feeder relay of Weather Radio broadcast program, nbfm
410.100	NOAA-to-local station feeder relay of Weather Radio broadcast program, nbfm
410.575	NOAA-to-local station feeder relay of Weather Radio broadcast program, nbfm
416.375	NOAA-to-local station feeder relay of Weather Radio broadcast program, nbfm
420.00-450.00	amateur radio, 70-cm uhf radio-frequency band in the electromagnetic spectrum
440.00-450.00	amateur radio, National Weather Service local and regional SKYWARN nets, nbfm
440.00-450.00	amateur radio, repeaters, local and regional communications, voice and packet, nbfm
449.000	NOAA wind-profiler radar to detect wind shear, 200 planned across U.S. in future
450.05-450.925	local broadcast station two-way communications
451.00-452.000	businesses and industries, local-coverage low-power transmitters, nbfm
451.78-452.014	businesses and industries, local and regional-coverage high-power transmitters, nbfm
452.325-452.95	local and regional railroad communications, nbfm
452.525-452.60	tow trucks, automobile clubs, nbfm
452.975-453.00	local newspaper communications, nbfm
453.025	highway emergency callboxes, nbfm
453.075	highway emergency callboxes, nbfm
453.125	highway emergency callboxes, nbfm
453.175	highway emergency callboxes, nbfm
454.000	oil spill cleanup, nbfm
455.05-455.925	local broadcast station two-way communications, nbfm
457.975-458.00	local newspaper communications, nbfm
460.000-470.00	meteorological satellites, WEFAX, telemetry, data
460.65-460.875	local airline operations, passenger problems, nbfm
460.90-460.975	local fire, ambulance and rescue service central alarm systems, nbfm
462.550	CB REACT teams, General Mobile Radio Svc (GMRS), repeater output channel 550
462.5625	CB REACT teams, General Mobile Radio Svc (GMRS), split simplex channel 5625
462.575	CB REACT teams, General Mobile Radio Svc (GMRS), repeater output channel 575
462.5875	CB REACT teams, General Mobile Radio Svc (GMRS), split simplex channel 5875
462.600	CB REACT teams, General Mobile Radio Svc (GMRS), repeater output channel 600
462.6125	CB REACT teams, General Mobile Radio Svc (GMRS), split simplex channel 6125
462.625	CB REACT teams, General Mobile Radio Svc (GMRS), repeater output channel 625
462.6375	CB REACT teams, General Mobile Radio Svc (GMRS), split simplex channel 6375
462.650	CB REACT teams, General Mobile Radio Svc (GMRS), repeater output channel 650
462.6625	CB REACT teams, General Mobile Radio Svc (GMRS), split simplex channel 6625
462.675	CB REACT teams, General Mobile Radio Svc (GMRS), repeater output channel 675
462.6875	CB REACT teams, General Mobile Radio Svc (GMRS), split simplex channel 6875
462.700	CB REACT teams, General Mobile Radio Svc (GMRS), repeater output channel 700
462.7125	CB REACT teams, General Mobile Radio Svc (GMRS), split simplex channel 7125
462.725	CB REACT teams, General Mobile Radio Svc (GMRS), repeater output channel 725
462.90-462.975	local emergency medical service (ems) communications
464.075-464.10	businesses and industries, local-coverage low-power transmitters
467.550	CB REACT teams, General Mobile Radio Svc (GMRS), repeater input channel 550
467.575	CB REACT teams, General Mobile Radio Svc (GMRS), repeater input channel 575
467.600	CB REACT teams, General Mobile Radio Svc (GMRS), repeater input channel 600

Frequency	Description
467.625	CB REACT teams, General Mobile Radio Svc (GMRS), repeater input channel 625
467.650	CB REACT teams, General Mobile Radio Svc (GMRS), repeater input channel 650
467.675	CB REACT teams, General Mobile Radio Svc (GMRS), repeater input channel 675
467.700	CB REACT teams, General Mobile Radio Svc (GMRS), repeater input channel 700
467.725	CB REACT teams, General Mobile Radio Svc (GMRS), repeater input channel 725
467.75-467.925	local businesses and chain stores
468.00-468.175	local emergency medical service (ems) communications
468.8250	NOAA GOES-West satellites used by Nat'l Bureau of Standards to relay time signals
468.8375	NOAA GOES-East satellites used by Nat'l Bureau of Standards to relay time signals
475.750	TV audio, U.S. television broadcast channel 14
481.750	TV audio, U.S. television broadcast channel 15
487.750	TV audio, U.S. television broadcast channel 16
493.750	TV audio, U.S. television broadcast channel 17
499.750	TV audio, U.S. television broadcast channel 18
505.750	TV audio, U.S. television broadcast channel 19
511.750	TV audio, U.S. television broadcast channel 20
517.750	TV audio, U.S. television broadcast channel 21
523.750	TV audio, U.S. television broadcast channel 22
529.750	TV audio, U.S. television broadcast channel 23
535.750	TV audio, U.S. television broadcast channel 24
541.750	TV audio, U.S. television broadcast channel 25
547.750	TV audio, U.S. television broadcast channel 26
553.750	TV audio, U.S. television broadcast channel 27
559.750	TV audio, U.S. television broadcast channel 28
565.750	TV audio, U.S. television broadcast channel 29
571.750	TV audio, U.S. television broadcast channel 30
577.750	TV audio, U.S. television broadcast channel 31
583.750	TV audio, U.S. television broadcast channel 32
589.750	TV audio, U.S. television broadcast channel 33
595.750	TV audio, U.S. television broadcast channel 34
601.750	TV audio, U.S. television broadcast channel 35
607.750	TV audio, U.S. television broadcast channel 36
613.750	TV audio, U.S. television broadcast channel 37
619.750	TV audio, U.S. television broadcast channel 38
625.750	TV audio, U.S. television broadcast channel 39
631.750	TV audio, U.S. television broadcast channel 40
637.750	TV audio, U.S. television broadcast channel 41
643.750	TV audio, U.S. television broadcast channel 42
649.750	TV audio, U.S. television broadcast channel 43
655.750	TV audio, U.S. television broadcast channel 44
661.750	TV audio, U.S. television broadcast channel 45
667.750	TV audio, U.S. television broadcast channel 46
673.750	TV audio, U.S. television broadcast channel 47
679.750	TV audio, U.S. television broadcast channel 48
685.750	TV audio, U.S. television broadcast channel 49
691.750	TV audio, U.S. television broadcast channel 50
697.750	TV audio, U.S. television broadcast channel 51
703.750	TV audio, U.S. television broadcast channel 52
709.750	TV audio, U.S. television broadcast channel 53
715.750	TV audio, U.S. television broadcast channel 54
721.750	TV audio, U.S. television broadcast channel 55
727.750	TV audio, U.S. television broadcast channel 56
733.750	TV audio, U.S. television broadcast channel 57
739.750	TV audio, U.S. television broadcast channel 58
745.750	TV audio, U.S. television broadcast channel 59

Frequency	Description
751.750	TV audio, U.S. television broadcast channel 60
757.750	TV audio, U.S. television broadcast channel 61
763.750	TV audio, U.S. television broadcast channel 62
769.750	TV audio, U.S. television broadcast channel 63
775.750	TV audio, U.S. television broadcast channel 64
781.750	TV audio, U.S. television broadcast channel 65
787.750	TV audio, U.S. television broadcast channel 66
793.750	TV audio, U.S. television broadcast channel 67
799.750	TV audio, U.S. television broadcast channel 68
805.750	TV audio, U.S. television broadcast channel 69
811.750	TV audio, U.S. television broadcast channel 70
817.750	TV audio, U.S. television broadcast channel 71
823.750	TV audio, U.S. television broadcast channel 72
829.750	TV audio, U.S. television broadcast channel 73
835.750	TV audio, U.S. television broadcast channel 74
841.750	TV audio, U.S. television broadcast channel 75
847.750	TV audio, U.S. television broadcast channel 76
853.750	TV audio, U.S. television broadcast channel 77
859.750	TV audio, U.S. television broadcast channel 78
865.750	TV audio, U.S. television broadcast channel 79
866.012-868.99	local public safety, mutual aid, fire, ambulance, rescue, police, nbfm
871.750	TV audio, U.S. television broadcast channel 80
877.750	TV audio, U.S. television broadcast channel 81
883.750	TV audio, U.S. television broadcast channel 82
889.750	TV audio, U.S. television broadcast channel 83
902.00-928.00	amateur radio, uhf radio-frequency band, local communications, repeaters, nbfm
1000.000	NOAA wind-profiler radar to detect very low wind shear, planned for future
1215.00-1300.00	amateur radio, 23-cm uhf radio-freq band, local communications, repeaters, nbfm
1270.00-1295.00	amateur television, amateur radio 23-cm band, local communications
1294.500	amateur radio, 23-cm band national simplex calling frequency, local communications
1525.0-1530.0	general satellite downlinks
1530.0-1544.0	maritime satellites
1614.000	GOES weather satellite-to-ground data
1626.500-1645.5	uplink to maritime satellites
1646.60-1710.00	WEFAX from weather satellites
1691.000	WEFAX from GOES stationary weather satellites, U.S.
1694.000	WEFAX from Meteosat weather satellite, Europe
1698.000	WEFAX from NOAA 6, NOAA 8 weather satellites
1702.500	WEFAX from NOAA 6, NOAA 9 weather satellites
2206.000	WEFAX from Spot, Earth-photography satellite, France
2209.086	WEFAX from GOES weather satellites
2211.000	WEFAX from Nimbus 7 weather satellite
2214.000	WEFAX from GOES weather satellites
2265.500	WEFAX from Landsat, Earth-photography satellite, USA
2273.500	WEFAX from Nimbus 7 weather satellite
2287.500	WEFAX from Landsat Earth-photography satellite, USA
7250.000-7750.0	WEFAX from weather satellites
8150.000	WEFAX from JERS-1, Earth-observation satellite, Japan
8175.00-8400.00	weather satellites downlinks
8350.000	WEFAX from JERS-1, Earth-observation satellite, Japan
8400.000	WEFAX from Spot, Earth-photography satellite, France
17200.0-17300.0	Earth exploration satellites
20200.0-21200.0	standard time signal satellites
24050.0-24250.0	government Earth exploration satellites
25250.0-27000.0	government standard time signal satellites

Abbreviations:

AFB	air force base
am	amplitude modulation, also a.m., a voice mode of transmission
amtor	similar to rtty on hf; similar to packet, but on lower long-distance hf frequencies
ARRL	American Radio Relay League
atv	amateur television, visual mode of communications
av	aviation
C-band	4,000-8,000 MHz or 4-8 GHz, radio-frequency band in the electromagnetic spectrum
CAP	Civil Air Patrol, volunteers searching for downed aircraft
common	used by all U.S. Coast Guard facilities
comms	communications
COSPAS	Russian acronym for Space System for Search of Vehicles in Distress, SARSAT/COSPAS
cw	continuous wave, International Morse code
data	digital computer information, such as the 50 megabits/second
DCN	U.S. Coast Guard Disaster Communications Network
digital	data transmissions
DOT	U.S. Department of Transportation
downlink	transmission from space
dpx	duplex communications, transmitting and receiving stations on different frequencies
ECN	U.S. Coast Guard emergency command network for emergencies or high priority traffic
ehf	extremely high frequency, 25,000-300,000 MHz or 25-300 GHz, radio-frequency band
ELT	Emergency Locator Transmitter
EPIRB	Emergency Position Indicating Radio Beacons
FAA	Federal Aviation Administration
fax	facsimile, on hf, usually 1.9 KHz above listed carrier frequency
FEMA	Federal Emergency Management Agency
FHWA	Federal Highway Administration, Department of Transportation
fm	frequency modulation, fm or wbfm or nbfm, a voice mode of transmission
freq	frequency, measured in cycles per second and labeled hertz
fsk	frequency shift keying in radioteletype or rtty, a teleprinter mode of communication
fstv	fast-scan television, a visual mode of communication
GHz	gigahertz, one billion hertz or one thousand megahertz or one million kilohertz
GSA	General Services Administration
ham	amateur radio operator
hertz	cycle per second
hf	high frequency, 3-30 MHz or 3,000-30,000 KHz, shortwave radio band
ht	handie-talkie, walkie-talkie, hand-held battery-powered portable radio
IB	Business, FCC land mobile radio service
IF	Forest Products, FCC land mobile radio service
IMO	International Maritime Organization
IP	Petroleum, FCC land mobile radio service
IS	Special Industrial, FCC land mobile radio service
IW	Power, FCC land mobile radio service
K-band	18,000-27,000 MHz, 18-27 GHz, radio-frequency band of the electromagnetic spectrum
Ka-band	27,000-40,000 MHz, 27-40 GHz, radio-frequency band of the electromagnetic spectrum
KHz	kilohertz, one thousand hertz
Ku-band	12,000-18,000 MHz, 12-18 GHz, radio-frequency band of the electromagnetic spectrum
L-band	1,000-2,000 MHz, 1-2 GHz, radio-frequency band of the electromagnetic spectrum
lf	low frequency, 300-3,000 KHz or .3-3 MHz, radio frequency band
LR	Railroad, FCC land mobile radio service
lsb	lower sideband, a voice transmission mode
lw	long wave, 30-300 KHz or 0.03-.3 MHz, radio frequency band
MARS	Military Affiliate Radio System, amateur radio service to Defense Department
mcw	modulated cw, Morse code can be heard on an a.m.-only radio
MHz	megahertz, one million hertz or one thousand kilohertz
millimeter band	40,000-300,000 MHz, 40-300 GHz, radio-frequency band of the electromag. spectrum
MUX	multiplexed
mw	medium wave, 300-3,000 KHz or .3-3 MHz, radio frequency band
NASA	National Aeronautics and Space Administration
NAVTEX	Int'l Maritime Organization worldwide automated broadcast system, warnings to mariners
nbfm	narrow band frequency modulation

NMA	main U.S. Coast Guard Atlantic Ocean area communication facility at Miami Florida
NMC	main U.S. Coast Guard Pacific Ocean area communication facility, San Francisco, Ca.
NMF	main U.S. Coast Guard Atlantic Ocean area communication facility at Boston, Mass.
NMG	main U.S. Coast Guard Atlantic Ocean area communication facility, New Orleans, La.
NMN	main U.S. Coast Guard Atlantic Ocean area communication facility, Portsmouth, Va.
NMO	main U.S. Coast Guard Pacific Ocean area communication facility at Honolulu Hawaii
NMR	main U.S. Coast Guard Atlantic Ocean area communication facility at San Juan, P.R.
NOAA	National Oceanic and Atmospheric Administration
NOJ	main U.S. Coast Guard Pacific Ocean area communication facility at Kodiak, Alaska
NORAD	North American Air Defense Command
NOZ	U.S. Coast Guard Atlantic Ocean area radio station at Elizabeth City, North Carolina
NRV	main U.S. Coast Guard Pacific Ocean area communication facility at Barrigada, Guam
NTS	National Traffic System, amateur radio operators forwarding radiograms
pkt	packet radio, computer-to-computer communications by radio
pm	phase modulation
psk	phase shift keying
pulse	usually transmissions by Omega or Loran-C navigational signals
QSO	conversation, a Morse code Q signal
rcpsk	raised cosine phase shift keying
rcv	receive
recon	reconnaissance
rpts	reports
RR	railroad
rtty	radioteletype, teleprinter comms, usually 170 or 850 shift, 2 KHz above carrier freq.
rv	recreational vehicle
S-band	2,000-4,000 MHz, 2-4 GHz, microwave radio-frequency band of the spectrum
sar	search and rescue
SAREX	Shuttle Amateur Radio Experiment, also CAP search and rescue exercise
SARSAT	Search and Rescue Satellite-Aided Tracking System, SARSAT/COSPAS
SHARES	nationwide federal emergency communications network for agencies
shf	super high frequency, 3,000-25,000 MHz, 3-25 GHz, radio-frequency band
shortwave	high frequency, hf, 3-30 MHz, radio-frequency band in the electromagnetic spectrum
sitor	teleprinter mode of communication
spx	simplex communications, transmitting and receiving stations on same frequency
ss	spread spectrum
ssb	single sideband, a voice transmission mode, occupying less bandwidth than am or fm
sstv	slow-scan television, visual comm like atv but slower and used on lower-freq. hf bands
stn	station
submillimeter	above 300,000 MHz, above 300 GHz, radio-frequency band of the electromg. spectrum
sw	shortwave, high frequency, hf, 3-30 MHz, 3-30 MHz, a radio-frequency band
swl	shortwave listener
tfc	traffic
tlm	telemetry
tv	television, a visual mode of communication
two meters	144-148 MHz, popular amateur radio frequency band in the electromagnetic spectrum
uhf	ultra high frequency, 300-3,000 MHz, 0.3-1 GHz, radio-frequency band
ulf	ultra low frequency, below 30 KHz, radio band in the electromagnetic spectrum
uplink	transmission toward space
USAF	United States Air Force
usb	upper sideband, a voice transmission mode
USCG	U.S. Coast Guard
USN	United States Navy
utc	coordinated universal time, z
vhf	very high frequency, 30-300 MHz, a radio-frequency band of the electromag. spectrum
VIP	very important person
vlf	very low frequency, 30-300 KHz or 0.03-.3 MHz, radio frequency band
wbfm	wide band frequency modulation
wx	weather
X-band	8,000-12,000 MHz, 8-12 GHz, radio-frequency band of the electromagnetic spectrum
xmit	transmit
Z	utc, coordinated universal time

Wind Chill Index

Temperature and wind cause heat to be lost from body surfaces. Cold and wind combine to make a body feel colder than the actual temperature.

For example, a temperature of 20 degrees Fahrenheit combined with a 20 mph wind causes body heat loss equal to that in –10 degrees with no wind. In another example, a calm-air temperature of 0 degrees Fahrenheit has an equivalent cooling effect of –39 degrees at a wind speed of 20 mph.

In the table below, WIND is wind speed in miles per hour. TEMP is temperature in degrees Fahrenheit. Speeds above 45 mph have little additional chilling effect.

WIND	0	5	10	15	20	25	30	35	40	45
TEMP	35	33	22	16	12	8	6	4	3	2
	30	27	16	9	4	1	–2	–4	–5	–6
	25	21	10	2	–3	–7	–10	–12	–13	–14
	20	16	3	–5	–10	–15	–18	–20	–21	–22
	15	12	–3	–11	–17	–22	–25	–27	–29	–30
	10	7	–9	–18	–24	–29	–33	–35	–37	–38
	5	0	–15	–25	–31	–36	–41	–43	–45	–46
	0	–5	–22	–31	–39	–44	–49	–52	–53	–54
	–5	–10	–27	–38	–46	–51	–56	–58	–60	–62
	–10	–15	–34	–45	–53	–59	–64	–67	–69	–70
	–15	–21	–40	–51	–60	–66	–71	–74	–76	–78
	–20	–26	–46	–58	–67	–74	–79	–82	–84	–85
	–25	–31	–52	–65	–74	–81	–86	–89	–92	–93
	–30	–36	–58	–72	–81	–88	–93	–97	–100	–102
	–35	–42	–64	–78	–88	–96	–101	–105	–107	–109
	–40	–47	–71	–85	–95	–103	–109	–113	–115	–117
	–45	–52	–77	–92	–103	–110	–116	–120	–123	–125

Winter Warning Words

blizzard	sustained wind speeds of 35 mph or more, steady snowfall, blowing snow
freezing rain or drizzle	ice coating is expected
heavy	weight can cause extra damage
snow	steady fall, unless occasional or intermittent is used
snow, intermittent	not a steady fall
snow, occasional	not a steady fall
snow flurries	intermittent snowfall, may reduce visibility
snow, heavy	4-6 in. or more in 12 hours, 6 or more in 24 hours, 2-3 where snow infrequent
snow, blowing	strong winds, poor visibility for lengthy period
snow, drifting	strong winds, poor visibility for lengthy period
snow squalls	brief intense snowfall with gusty winds
watch	winter storm approaching
warning	winter storm imminent

Heat Index

High humidity and high temperature reduce a body's ability to cool itself. The Heat Index is a measure of what hot weather feels like to an average person. For example, when the actual air temperature is 100 degrees Fahrenheit and the relative humidity is 50 percent, the effect on a human body would be the same as at 120 degrees. Heat exhaustion and sunstroke are likely when the Heat Index reaches 105 degrees. In another example, an air temperature of 85 degrees and a relative humidity of 90 percent combine to make an apparent temperature of 102 degrees.

In the table, TEMP is actual air temperature in degrees Fahrenheit. HUMID is percentage of relative humidity. The body of the table shows apparent temperature.

TEMP	70	75	80	85	90	95	100	105	110	115	120
HUMID											
0%	64	69	73	78	83	87	91	95	99	103	107
10%	65	70	75	80	85	90	95	100	105	111	116
20%	66	72	77	82	87	93	99	105	112	120	130
30%	67	73	78	84	90	96	104	113	123	135	148
40%	68	74	79	86	93	101	110	123	137	151	
50%	69	75	81	88	96	107	120	135	150		
60%	70	76	82	90	100	114	132	149			
70%	70	77	85	93	106	124	144				
80%	71	78	86	97	113	136					
90%	71	79	88	102	122						
100%	72	80	91	108							

Summary Warning Words

watch	threatening weather is most likely to occur during time period; watch and plan.
warning	threatening weather actually has been seen, take safety precautions.
small craft advisory	two hours or more of hazardous marine weather or sea conditions; often wind speeds of 21 to 38 mph with dangerous waves.
gale warning	wind speeds of 39 to 73 mph, not associated with a tropical storm.
storm warning	wind of 55 mph and above, not associated with a tropical storm.
tropical storm warning	coastal area winds of 39 to 73 mph in a 24-hour period.
special marine warning	two hours or less of winds of 39 mph or more with dangerous waves.
cyclone	tornadoes, hurricanes, lows shown on weather maps are examples of cyclones.
thunderstorm, severe	thunderstorm with winds of 58 mph and/or 0.75 inch or larger hailstones.
tornado	violent rotating column of air contacting ground.
tornado watch	severe tornado-spawning thunderstorm is likely; plan ahead; listen for warning.
tornado warning	tornado has been seen by radar; take safety precautions now.
hurricane	severe cyclone originating over warm tropical waters, winds 74 mph or higher.
hurricane watch	hurricane threatens coastal and inland communities.
hurricane warning	hurricane-force winds of 74 mph and above, dangerous waves and extremely high water are expected within 24 hours; associated with a tropical storm.
typhoon	severe cyclone originating over warm tropical waters, winds 74 mph or higher.
flood	water overflows natural or artificial banks of stream, river, or lake, or accumulates by drainage over low-lying areas.
nautical/statute mile	one nautical mile = 1.1516 statute mile; one statute mile = 0.8684 nautical mi.

Index

AAA
Abadan VOLMET, 42
active antenna, 28
Adelaide VOLMET, 41
Aden, Saudi Arabia, 11
AFB, 108
AFOS, HHAPP, packet radio, 38
Africa, 9, 11, 45
Ahmadabad VOLMET, 42
air band, 43
air force base, 108
air masses, 6
aircraft emergency, 42
aircraft, WP-3D, 37
airport weather, 43
Alaska, Fairbanks, 10
    Kodiak, 109
alerting tones, 13
Alexandria, Virginia, 48
Alger VOLMET, 42
Algiers VOLMET, 41
Alice Springs VOLMET, 41
altitude measurements, 5
a.m., 108
    broadcast band, 53
Amateur Radio Emergency Svc, 54, 56
amateur radio, 54
    activities, 54
    ARES, 54, 56
    bands, 55
    bulletins, W1AW, 86
    club, 56
    emergencies, 54
    frequency bands, 55
    net frequencies, 54, 56
    nets, SWL, 86
    operator, 108
    packet, 37
    RACES, 56
    repeaters, 54
    transceivers, 13
    weather crises, 54
    wind profilers, 86
amateur television, 55, 108
American Radio Relay Leag., 54, 87, 108
Amman VOLMET, 42
amplitude modulation, 55, 108
AMSAT, 24, 29
Amsterdam VOLMET, 41
amtor, 55, 108
analog-to-digital, 30
ANARC, 87
Anchorage VOLMET, 41
Anguilla, 53
Ankara VOLMET, 41, 42
Annaba VOLMET, 42
annual cycle, 10
Antarctica, 9, 10, 11
antenna, radio, 27
    active, 28
    beam, 28
    discone, 28
    omni-directional, 28
    radio, WEFAX, 28
    turnstile, 28
anticyclone, 6
APT, 22, 30
arctic, 10
ARES, 54, 56
    RACES, 56
Argentina, 9
Ariane rocket, 23
Arica, Chile, 11
Aristotle, 5
Arkansas, 87
Army forecasts, 6
ARRL, 54, 108
    station W1AW, 87

Arthur C. Clarke, 17
Asia, 9, 11, 45
Assn No. American Radio Clubs, 87
Astrakhan, Russia, 11
astronomy, 5
AT&T high-seas radiotelephone, 44, 46
Athens VOLMET, 41, 42
ATIS, 43
Atlantic Ocean, 12, 47, 109
    hurricane, 37
    launches, 17
Atlantic Ocean region VOLMET, 42
atmosphere, low levels, 6
    state of, 10
atmospheric circulation, 5
    motions, 6
    patterns, 8
    transformations, 6
ATS-3, satellite, 37
attitude-control system, 15
atv, 55, 108
Auckland VOLMET, 41
Australia, 9, 11, 27
automated systems, 7
    Automated Terminal Info Svc, 43
    Automated Weather Observ Sys, 43
aviation, 108
aviation weather, 41
    VOLMET, 41
AWOS, 43
    content pattern, 43
Azizia, Libya, 9

BBB
B-52 bomber, 17
Baghdad VOLMET, 42
Bahrain VOLMET, 42
Baikonur Cosmodrome, 17
balloons, 6
Baltimore, Maryland, 47
bands of frequencies, 88
bandwidth filter, i.f., 29
Bangkok VOLMET, 41
barometer, 5
Barrigada, Guam, 109
Basrah VOLMET, 42
Batagues, Mexico, 11
bbs, 30, 55
    Climate Assessment, 13
    East Coast Marine Users, 13
    NOAA, Boulder, Colorado, 13
beam antenna, 28
Beirut VOLMET, 42
Belgium, 21
Ben Gurion VOLMET, 41
Benghazi VOLMET, 42
Berlin VOLMET, 41
Big Sur, 20
Billings, Montana, 10
Bjerknes, Wilhelm, 6
    Jacob, 6
blizzard, 12, 110
Boise, Idaho, 10
Bombay VOLMET, 42
Bordeaux VOLMET, 41
Boston, Massachusetts, 47, 109
Boulder, Colorado, 7
Boyle, Robert, 5
Bracknell, Great Britain, 8
Bratislava VOLMET, 41
Brazil, 24
Brisbane VOLMET, 41
British Columbia, 11
British West Indies, 53
Brno VOLMET, 41
broadband pre-amp, 29
broadcast band, a.m., 53
Budapest VOLMET, 41

bulletin boards, 30, 55
    Climate Assessment, 13
    East Coast Marine Users, 13
    NOAA, Boulder, Colorado, 13
buoys, 7
business, land mobile radio svc, 52, 108

CCC
C-band, 88, 108
Cairo VOLMET, 42
Calcutta VOLMET, 42
California, 9
    Fresno, 10
    Grant Park, 10
    Redwod City, 20
    San Francisco, 109
    Vandenberg AFB, 17, 20
Cameroon, 11
Camp Springs, Maryland, 6
Canada, 9, 11, 23, 24, 25, 48, 53, 54
CAP, 56, 108
    search and rescue exercise, 109
Cape Canaveral, Florida, 17
Cape Cod, 26
Cape Verde Islands, 51
Caribbean Sea, 12
    hurricane, 37
category 1-to-5, hurricane, 38
CB transceivers, 13
CDA, 19
Central America, 45
Challenger disaster, 23
Charlotte Pass, New South Wales, 9
Charney, Jule, 7
Chernobyl, 24
Cherrapunji, India, 11
Cheung Chau VOLMET, 41
Chile, 10, 11
chilling effect, 110
China, 9, 23, 27, 30
    weather satellites, 22
Chitose VOLMET, 41
chlorophyll content, 20
circular orbit, satellite, 17
    periods, 17
Civil Air Patrol, 56, 108, 109
Clarke Belt, 17
Clarke, Arthur C., 17
Climate Analysis Center, 8
Climate Assessment bbs, 13
climate, 10
    maps, 20
climatologists, 8
climatology, 6
clocks, recording, 5
Cloncurry, Queensland, 9
cloud cover photos, 35
    first satellite photos, 14
cloud formations, 14
    classification, 5
    radar, 7
cloudiness, 10
US Coast Guard, 27, 42, 47, 108, 109, 110
coastal stations, 14, 45
coaxial cable, 31
codes, VORTEX, 38
    RECCO, 39
Cold Bay VOLMET, 41
Colombia, 11
Colorado, 87
    Boulder, 7
    Fort Collins, 48
Command & Data Acq Stn, 19
common, 108
communications, 108
CompuServe, 30
computer, 27, 33, 51
    forecasting, 7

computer, printer, 27
computer-to-computer radio, 109
Connecticut, 87
    Newington, 54
continental, 10
continuous wave, 47, 54, 55, 108
convective outlook, 12
converter, frequency, 28
cooling, law, 5
Coordinated Universal Time, 7, 56, 110
Copenhagen VOLMET, 41
Coral Gables, Florida, 6, 12, 37
Corpus Christi, Texas, 47
Cosmos 4, satellite, 15, 21
Cosmos 1766, satellite, 21
Cosmos 1861, satellite, 25
COSPAS, 24, 27, 108
Crimean War, 5
crises, amateur radio, 54
Crkvica, Yugoslavia, 11
crop health, 20
cup anemometer, 5
cw, 47, 54, 55, 87, 108
cycles per second, 88, 108
cycle, annual, 10
    diurnal, 10
cyclones, 8, 110
    theory, 6

DDD

d'Alembert, Jean Le Rond, 5
D-layer, ionosphere, 44
daily-range forecasts, 8
Dakar VOLMET, 42
Dalton, John, 5
Darmstadt, Germany, ESA, 21
Darwin VOLMET, 41
data, 108
    transmissions, 108
DCN, 108
Death Valley, California, 9
Debundscha, Cameroon, 11
DeCartes, Rene, 5
Defense Meteorological Satellite, 20
Delhi VOLMET, 41
demodulator, 30
    FSK, 51
Denmark, 21
Department of Transportation, 108
depression number, 37
Dept. of Agriculture, 6
Dept. of Commerce, 6
desert, remote areas, 14
deviation, fm, 29
Dhahran VOLMET, 42
Diego Suarez VOLMET, 42
digipeater, 55
digital, 108
    computer, 108
    forecasting, 7
Disaster Communications Network, 108
disaster potential, hurricane, 38
discone antenna, 28
distance, 56
diurnal cycle, 10
Doppler radar, 7, 87
Doppler shift, 25
DOT, 108, 109
downlink, 108
DRIG, 30
droplets, suspended, 7
drought, 8
ducting, 45
duplex communications, 108
DX, 56
    tips, 87
dynamic equations, 6, 7

EEE

E-layer, ionosphere, 44
Earth Observation Satellite Co., 23
Earth Radiation Budget, 16
Earth's atmosphere, 5
Earth-observing satellites, 23
Earth-resources satellites, 23

East Coast Marine Users bbs, 13
echoes, radar, 7
ECN, 108
Ecuador, 27
Edmonton, 42
EDS, 6
ehf, 88, 108
Eisenhower, Dwight D., 14
El Nino, 21
Elat VOLMET, 41
electric instruments, 5
electromagnetic energy spct, 88, 108, 109
elevation, 11
Elizabeth City, North Carolina, 109
elliptical orbit, satellite, 17
ELT, 25, 108
emergency, 56
    aircraft, 42
    amateur radio, 54
    bulletins, W1AW, 86
    messages, 54
    net frequencies, 54
    preparedness, 54
Emergency Locator Transmitter, 25, 108
Emergency Position Ind. Radio Bcn., 108
empirical rules, 5
energy spectrum, 88
England, radio hobbyists, 21
Environmental Data Svc, 6
Environmental Sci Svcs Admin, 6
Eosat, 23
EPIRB, 25, 27, 108
equatorial orbits, 17
equipment manufacturers, 31
ERBE, 16
ESA, 21, 22
Esperanza, Palmer Peninsula, 9
ESSA, 6
    weather satellites 15
Europe, 9, 11, 23, 45
Europe region VOLMET, 41
European Center, Bracknell, 8, 9
European Space Agency, 17, 21, 22, 24
European weather satellites, 21
evaporation, 5
Explorer satellites, 14
    Explorer 6, 14
extended forecasts, 8
extrapolation, 8
extremely high frequency, 108
extremes, 9
    highest temperatures, 9
    least precipitation, 11
    lowest temperatures, 9
    most precipitation, 11

FFF

F1-layer, ionosphere, 44
F2-layer, ionosphere, 44
FAA, 42, 108, 109
    weather, 42
facsimile, 55, 108
    radio, 30
    weather, 22
Fahrenheit, 10, 110
Fairbanks VOLMET, 41
Fairbanks, Alaska, 10
fast-scan television, 54, 55, 108
fax, 33, 55, 108
    radio, 27, 30
FCC, 52
FDM, 30
FEC, 48
Federal Aviation Administration, 108
Federal Communications Commission, 52
Federal Emergency Mngmnt Agency, 108
Federal Highway Administration, 108
FEMA, 108, 109
Feng Yun, satellite, 22
FHWA, 108, 109
Field Day, 54
filter, i.f. bandwidth, 29
Finland, 21
fishing and shipping, 14
fishing fleet, Russian, weather, 51
fix, location, 25

fleet, Russian, weather, 51
flood, 110
Florida, 10, 12
    Cape Canaveral, 17
    Coral Gables, 6, 37
    Miami, 109
fm, 29, 55, 87, 108
    deviation, 29
    narrow-band, 29
fog, 14, 20
forecasting, 5
    computers, 7
    weaknesses, 8
forest fires, 20
forest prdt, land mobile radio svc, 52, 108
Fort Collins, Colorado, 48
France, 21, 24, 25
    Spot satellite, 23
Franklin, Benjamin, 5
freezing rain, 110
French fleet, 5
French Guiana, 24
    launches, 17
frequencies, 108
    amateur radio, nets, 54, 56
    ATIS, 43
    AWOS, 43
    bands, 88
        amateur radio, 55
    Hurricane Hunters, 37
    Land Mobile Radio Svc, 52
    observation satellite, 24
    popular weather satellite, 22
    shortwave WEFAX, 33
    time stations, 49
    Traveler's Information Svc, 52
    updates, SWL nets, 86
    VOLMET, 41
    Weather Radio, 13
    weather satellites, 22
    weather, 87, 88
    wind profilers, 86
frequency, 108, 108
    converter, 28
    emergency, aircraft, 42
    explained, 88
    kilohertz, 55
    megahertz, 55
    updates, 87
frequency division multiplexing, 30
frequency modulation, 29, 108
frequency shift keying, 108
Fresno, California, 10
fronts, 6, 20
fsk, 108
    demodulator, 51
fstv, 55, 108
Fukuoka VOLMET, 41

GGG

GaAs-FET pre-amplifier, 29
gale warning, 110
Galileo Galilei, 5
Galveston, Texas, 47
gamma rays, 88
Gander VOLMET, 42
gases, study, 5
Gaya VOLMET, 41
General Electric Co., 23
General Services Administration, 108
General Weather Service, 6
geodetic surveys, 6
geostationary, 17, 18
Germany, 21
GHz, 88, 108
gigahertz, 88, 108
global telecommunications system, 7
GMT, 7, 56, 110
Goddard Space Flight Ctr, 24
GOES, 18, 28, 17, 25, 30, 31, 35, 37, 51
    GOES-NEXT, 19
    launches, 20
    shortage, 19
GPS, 48
Grand Banks, 47
Grand Haven, Michigan, 10

Grant Park, California, 10
graphs, weather, 7
gray-scale, 35
Great Britain, 21, 25
    Bracknell, 8
Great Plains, 11
Greece, 21
Greek, meteoron, 5
Greenhouse Effect, 16
Greenland, 9, 47
Greenwich Mean Time, 7, 56, 110
ground wave, 45
GSA, 108
GTO, 17
Guam VOLMET, 41
Guam, Barrigada, 109
Gulf of Mexico, 11, 12
Gulf of St. Lawrence, 14

HHH

Hadley, John, 5
Hagerstown, Maryland, 53
Haleakala Summit, Maui, Hawaii, 9
Halifax, 42
Halley, Edmund, 5
ham radio, 54, 108
    bulletins, W1AW, 86
    emergencies, 54
    emergency nets, 54
    net frequencies, 56
    SWL nets, 86
    weather crises, 54
    wind profilers, 86
handie-talkie, 56, 108
hardware manufacturers, 31
Hawaii, 9, 10, 11, 27
    Honolulu, 109
    Kauai, 48
heat exhaustion, 110
heat index, 110
Henderson Lake, Canada, 11
Hertz, 88, 108
hf, 55, 88, 108, 109, 110
HHAPP, 37
high frequency, 55, 108, 109, 110
high-seas radiotelephone operator, 46
highest temperatures, 9
Hilo VOLMET, 41
Himawari, Japan, satellite, 22
Hong Kong VOLMET, 41
Honolulu Hawaii, 109
Honolulu VOMET, 41
Hooke, Robert, 5
Hope, satellite, 26
horizon scanner, 14
Howard, Luke, 5
ht, 56, 108
Hughes Aircraft Co., 19, 23
humidity, 5, 10, 110
hurricane, 8, 14
    disaster potential, 38
    season, 19
    warning, 110
    watch, 110
hurricane hunters, 37
    aircraft, 12
    frequencies, 37
    identifying, 37
hydrological radio use, 52
hydrological stations, 6
Hydromet Service, Russia, 51
hygrometer, 5
hypsometry, 5

III

i.f. bandwidth filter, 29
IB, land mobile user, 52, 108
ICAO, 37
ice cover, 20
iceberg patrols, 47
ICOM, 28
Idaho, Boise, 10
identifying hurricane hunters, 37
IF, land mobile user, 52, 108
Ifrane, Morocco, 9

IMO, 108
Improved TIROS, 16
Independencia, freighter, 26
index, heat, 110
    wind chill, 110
India, 11, 22, 23
    ISRO, 23
    weather satellites, 22
Indian Ocean region VOLMET, 42
Indian Space Rresearch Org, 23
industrial activities, 12
infrared, telescope, 19
    light, 88
    photos, 14
    radiometer, 15, 20
Insat, satellite, 22
intl satellite cloud project, 8
interface, radio, 27
intermediate frequency, 29
International Maritime Organ., 108, 109
International Morse code, 47, 54, 55, 108
Intl Meteorological Organization, 8
ionosphere, 44
    D-layer, 44
    E-layer, 44
    F1-layer, 44
    F2-layer, 44
    Kennelly-Heaviside Layer, 44
IP, land mobile user, 52, 108
IRAC, 87
Ireland, 21
IS, land mobile user, 52, 108
island weather stations, 14
Israel, 9
ISRO, 23
Istanbul VOLMET, 41, 42
Italy, 21
ITOS, 16
IW, land mobile user, 52, 108

JJJ

Japan weather satellites, 22
    JERS-1, satellite, 24
    Kagoshima Island, 22
    Kyushu, 22
    MOS, satellite, 23
    Tokyo, 22
Jeddah VOLMET, 42
JERS-1, Japan, satellite, 24
JESAUG, 30
jet stream, 6, 14, 20
Jiwani VOLMET, 41
journalists, satellite, 24

KKK

K-band, 88, 108
Ka-band, 88, 108
Kagoshima Island, Japan, 22
Kahului VOLMET, 41
Kansas City, Missouri, 6, 12
Kanton Island, 27
Kao Hsiung VOLMET, 41
Karachi VOLMET, 41
Kauai, Hawaii, 11, 48
Kazakhstan launches, 17
KCC, public coast station, 47
Kennelly-Heaviside Layer, ionosphere, 44
Kenyan, 27
Keplerian elements, 29
    Keps, 29
keyboard communication, 55
KHz, 55, 108
kick motor, 17
kilohertz, 88, 108
King Salmon VOLMET, 41
kites, 6
KLH, public coast station, 47
KMI, AT&T station, 46
Kodiak, Alaska, 109
Kota Kinabalu VOLMET, 42
Kourou launches, 17
KOW, public coast station, 47
KQP, public coast station, 47
Ku-band, 88, 108
Kuala Lumpur VOLMET, 42

Kuwait VOLMET, 42
Kyushu, 22

LLL

L-band, 88, 108
Labrador Current, 47
Lahr, 42
Lake Michigan, 10
Land Mobile Radio Service, 52
Landsat, satellite, 23
latitude, North America, 11
launch pad, 17
laws of cooling, 5
laws of motion, 5
laws of refraction, 5
lf, 55, 108
Libya, 9, 24
    desert, 10
lightning, 5
    detectors, 12
line-of-sight propagation, 46
local airport advisories, 42
local predicting, 9
location fix, 25
London Heathrow VOLMET, 41
longitude, North America, 11
longwave, 87, 108
Loran-C, 109
Los Angeles VOLMET, 41
Louisiana, New Orleans, 47, 109
low frequency, 55, 109
low-altitude satellites, 17, 22
lower sideband, 55, 109
lowest temperatures, 9
LR, land mobile user, 52
lsb, 55, 87, 109
lw, 109
Lyon VOLMET, 41

MMM

Macao VOLMET, 41
mackerel, 51
Madrid VOLMET, 41
magnetic storms, Sun, 44
Manila VOLMET, 41
marine weather, 44
    radiotelephone stations, 46
    transceivers, 13
Maritime Observation Satellite, 23
maritime, weather, 44
    bulletin service, NAVTEX, 48
    climate, 10
    weather services, 44
Marriotte, Edme, 5
MARS, 56, 109
Marseille VOLMET, 41
Marshall Space Flight Center, 16
Maryland, Baltimore, 47
    Camp Springs, 6, 26
    Hagerstown, 53
    Rockville, 6
    Suitland, 8, 19
Massachusetts, Boston, 47, 109
Maui, Hawaii, 9
Mayday, 25
mcw, 109
medium frequency, 45
medium-range forecasts, 8
medium-wave band, 53, 87, 109
megahertz, 88, 109
message codes, VORTEX, 38
messages, emergency, 54
Meteor, satellite, 17, 21, 22, 30, 31
Meteorologica, 5
meteorological radio use, 52
meteorological stations, 6
meteorologists, 8
meteorology, 5
    modern, 7
meteoron, 5
Meteosat, satellite, 21
Mexico, 11, 54
MHz, 55, 109
Miami, Florida, 109
Michigan, Grand Haven, 10

microwave, 88
    radio, 109
        sounder, 16
Middle East region VOLMET, 42
Midwest, 11
Military Affiliate Radio System, 56, 109
military satellites, 20
    weather, 42
millimeter band, 109
Milwaukee, Wisconsin, 10
Mir, 109
mission number, 37
Missouri, 12
    Kansas City, 6
modes of operation, 55
modulated cw, 109
moisture patterns, 5
monitor, radio, 27
monsoon, 8
Montana, Billings, 10
MOP, 21
Morocco, 9
Morse code, 47, 54, 56
    Q signal, 109
MOS, satellite, 23
Moscow VOLMET, 41
motion, law, 5
Mt. Waialeale, Kauai, Hawaii, 11
Mulka, South Australia, 11
multiplexed, 109
Munich VOLMET, 41
Murmansk, 51
MUX, 109
mw, 109

NNN

Nadezhda, Hope, satellite, 26
Nagoya VOLMET, 41
Naha Okinawa VOLMET, 41
Nantucket, 26
narrow band frequency mod., 29, 55, 109
NASA, 17, 18, 37, 109
NASDA, 22
Nat'l Hurricane Center, 12, 37
Natl Aeronautics & Space Admin, 109
National Traffic System, 56, 109
National Weather Service, 6, 12, 44, 56
Natl Bureau Standards, 51
Natl Hurricane Ctr, 6
Natl Inst Standards & Technology, 48
Natl Marine Fisheries Svc, 6
Natl Meteorological Ctr, 6, 8, 9
Natl Ocean Survey, 6
Natl Oceanic & Atmospheric Adm., 6, 109
Natl Severe Storms Forecast Ctr, 6
Natl Spac Devel Agency, Japan, 22
natural disasters, 13
nautical mile, 110
NAVAREA, 45
    warnings, 45
navigational signals, 109
    warnings, 109
Navstar, 48
NAVTEX, 47, 48, 48, 109
Nawabshah VOLMET, 41
nb, 109
nbfm, 108, 109
Nebraska, 87
NESDIS, 27, 33
Netherlands, 21
nets, 56
    emergency, amateur radio, 54
    SWL, 87
network, 56
New Guinea, 22
New Orleans, Louisiana, 47, 109
New South Wales, 9
New Tokyo VOLMET, 41
New York VOLMET, 41
New York, New York, 47
    Port Washington, 26
Newington, Connecticut, 54
Newton, Issac, 5
NEXRAD, 7
Next Generation Radar, 7
NHC, 12

Nice VOLMET, 41
Nicosia VOLMET, 42
Nimbus weather satellites, 15, 17, 18, 20
NIST, 48, 51
NMA, 109
NMC, 109
NMF, 109
NMG, 109
NMN, 109
NMO, 109
NMR, 109
NOAA, 6, 13, 18, 27, 33, 87, 109
    alert tones, 13
    bulletin boards, 13
    satellite, 16, 17, 20, 22, 31
        launches, 18
    Weather Radio, 13
    TIS, 53
node, 55
NOJ, 109
non-commercial radio service, 54
NORAD, 109
    Keps, 30
Norfolk, Virginia, 47
North Africa region VOLMET, 42
North America, 9, 10, 11
North American Air Defense Cmnd., 109
North Atlantic region VOLMET, 41
North Carolina, Elizabeth City, 109
Northern Cosmodrome, 17, 21
Northice, Greenland, 9
Norway, 21, 25
notice to mariners, 109
nowcasts, 8
NOZ, 109
NRV, 109
NTIA, 87
NTS, 56, 109
nuclear attack, 13
numerical models, 7
    prediction, 7
NWS, 6

OOO

Oakland VOLMET, 41
observation number, 37
observation satellite frequencies, 24
ocean, buoys, 7
    currents, 20
    currents, cold, 10
    temperatures, 20
    winds, 7
Oceania, 9, 11
oceanographic stations, 6
Okean, satellite, 22
Omega, 48, 109
omni-directional antenna, 28
ONSCEN, 48
Ontario VOLMET, 41
Operations Centre, ESA, 21
operator, high-seas, 46
Oran VOLMET, 42
orbit elements, Keplerian, 29
orbits, satellites, 17
Osaka VOLMET, 41
OSCAR-7, 24
Oslo VOLMET, 41
Oymkyon, Russia, 9
ozone levels, 20

PPP

Pacific islands, 10
Pacific Ocean, 11, 21, 27, 109
    launches, 17
Pacific region VOLMET, 41
packet radio, 55, 109
    bulletin board system, 55
    hurricane, 37
    network, 55
    repeater, 55
Pakistanis, 27
Palmer Peninsula, 9
Paris-Le Bourget VOLMET, 41
Paris-Orley VOLMET, 41
particle density, 16

Pascal, Blaise, 5
pbbs, 55
Pegasus rocket, 17
Penang VOLMET, 42
People's Republic of China, 22
personal computer, 27, 33, 51
petroleum, land mobile radio svc, 52, 108
phase modulation, 109
phase shift keying, 109
Philippines, 9
physics, laws, 6
Plesetsk, Russia, 21
    launches, 17
plots, weather, 7
pm, 109
polar air, maritime, 11
polar areas, 14
polar-front theory, 6
polar-orbit weather satellite, 15, 17, 27
pollution levels, 20
popcorn clouds, 14
Port Washington, New York, 26
Portland VOLMET, 41
Portsmouth, Virginia, 48, 109
Portugal, 21
potential hurricane disaster, 38
power, land mobile radio service, 52, 108
Prague VOLMET, 41
pre-amplifier, 28, 29
    radio, 27
precipitation, 10
    formation, 6
predicting weather, 8
pressure, 10
    anemometer, 5
Prestwick VOLMET, 41
printer, 33
    computer, 27
Private Land Mobile Radio Service, 52
profilers, wind, 86, 87
PROFS, 7
propagation, 53, 54
    ducting
    shortwave skip, 44, 88
psk, 109
Puako, Hawaii, 11
public coast stations, 44, 46
Puerto Rico, San Juan, 48, 109
pulse, 109

QQQ

Q codes, 56
Q signal, 109
QRT, 56
QRU, 56
QRV, 56
QRX, 56
QSL cards, 54, 56
QSO, 56, 109
QSY, 56
Queensland, 9, 11
Quibdo, Colombia, 11

RRR

RACES, 54, 56
radar, 7, 88
    Doppler, 7, 87
    weather, 21
Radio Amateur Civil Emerg. Svc, 54, 56
Radio Shack, 28
radio, 88
    bulletin board system, 55
    hobby, 54
    antenna, 27
        active, 28
        turnstile, 28
        beam, 28
        discone, 28
        omni-directional, 28
        WEFAX, 28
    bandwidth filter, 29
    broadband pre-amp, 29
    computer, 27
    facsimile, 27, 30
    fax, 27, 30

frequency converter, 28
GaAs-FET pre-amp, 29
interface, 27
lightweight, battery, 6
monitor, 27
monitoring severe weather, 12
NOAA, 13
radio, pre-amplifier, 27, 28, 29
receive, shortwave, 30
receiver, 29, 31
    shortwave, 33
reporting, 6
scanner, 22, 27, 29
severe weather, 12
shortwave receiver, 51
uhf receiver, 27
vhf receiver, 27
weather band, 13
WEFAX, 27
radio-frequency band, 109
radiograms, 109
radiometer, 16, 18
    infrared, 15, 20
    Meteosat, 21
    spin-scan, 19
radiosondes, 6
radiotelephone, 44
    stations, 46
radioteletype, 51, 54, 55, 108, 109
railroad, 109
    FCC land mobile radio svc, 52, 109
rain, 8, 20
raised cosine phase shift keying, 109
Ramat David VOLMET, 41
Rangoon VOLMET, 41
rbbs, 55
rcpsk, 109
Reagan, Ronald, 23
RECCO reporting codes, 39
    telemetry code, 39
receive shortwave WEFAX, 33
    weather satellites, 27
receiver, radio, 29, 31
    weather satellite, 29
reconnaissance, 109
recreational vehicle, 109
Redwood City, California, 20
refraction, law, 5
Reims VOLMET, 41
remote sensing, 7
repeater, 55
    amateur radio, 54
reporting codes, RECCO, 39
reports, 109
rescue, weather satellites, 24
Ribadavia, Argentina, 9
ridges, upper-level, 14
river levels, 20
robot ocean buoys, 7
Rockville, Maryland, 6
RS-10/11, satellites, 25
rtty, 55, 87, 108, 109, 109
Russia, 9, 11, 17, 30, 108
    fleet weather, 51
    Hydromet Service, 51
    SARSAT, 24
    satellite, 22

SSS

S-band, 88, 109
Saffir-Simpson Scale, 38
Saigon VOLMET, 41
Saint Denis VOLMET, 42
San Francisco, California, 109
    Bay, 10
San Francisco VOLMET, 41
San Juan, Puerto Rico, 48, 109
SAR, weather satellites, 24, 109
SAREX, 56, 109
Sarmiento, Argentina, 9
SARSAT, 24, 26, 27, 109
SARSAT/COSPAS, 108, 109
Satellite Ops Cntrl Ctr, 19
satellite, weather, 7, 14
    ATS-3, 37
    China, 22

Cosmos 4, 15
Earth-observing, 23
ESSA series, 15
Explorer 6, 14
first cloud TV, 14
frequencies, 22
GOES, 18, 28, 37
how to receive, 27
India, 22
Japan, 22
JERS-1, 24
journalists, 24
Landsat, 23
Nimbus series, 15
NOAA series, 16
observation, freqs, 24
orbits, 17
periods, circular orbits, 17
photos, infrared, 14
photos, visible-light, 14
polar orbit, 27
search and rescue, 24
SMS, 18
Spot, 23, 24
stationary orbit, 28
time signals, 51
TIROS series, 14
USSR, 15
Saudi Arabia, 11
scanner, radio, 22, 27, 29
scatterometer, 7
Scott AFB, Illinois, 26
sea ice, 14, 16
sea level, 10
search and rescue, 20, 109
    satellites, 24, 108, 109
    Search And Rescue, CAP, 56
    Search and Rescue Satellite-Aided
        Tracking System, 109
    search for extraterrestrial intel., 109
    weather satellites, 24
SEASAT, 23
seasonal changes, 10
Seattle VOLMET, 41
Seattle, Washington, 47
section, 56
Seoul VOLMET, 41
SETI, 109
Severe Storm Forecast Ctr, 12
severe thunderstorm, 110
severe weather warnings, 7
    weather, radio, 12
Seville, Spain, 9
Shannon VOLMET, 41
Shanxi province, China, 22
SHARES, 109
Shemya VOLMET, 41
shf, 88, 109
ship-to-shore, 44
short skip, 45
short-range forecasts, 8
short-term climate forecasts, 8
shortwave, 55, 87, 88, 109, 110
    frequencies, 44
    radio band, 108
    receiver, 30, 33, 51
        WEFAX, 33
    shortwave listener, 110
    shortwave listening nets, 87
    skip, propagation, 44
    SWL, 110
    WEFAX frequencies, 33
Shuttle Amateur Radio Exper., 56, 109
simplex, 109
Singapore VOLMET, 41
single sideband, 55, 109
SITOR, 48, 109
skip signals, 44, 88
    best time of year, 45
    short, 45
    shortwave propagation, 44
    zone, 44
sky-wave signals, 44
SKYWARN, 56
slow-scan television, 54, 55, 109
small craft advisory, 110
smart digi, 55

SMS weather satellites, 18
Snag, Yukon, Canada, 9
snow, 8, 110
snow cover, 14, 16, 20
SOCC, 19
software publishers, 31
solar cells, 14
solar disturbance, 16
solar energy, 11
solar flares, 45
soldering iron, 29
sounder, microwave, 16
    stratospheric, 16
    vertical atmospheric, 19
sourcelist, manufacturers, 31
South America, 9, 11, 45
South Australia, 11
South Pole Station, 11
Southeast Asia region VOLMET, 41
Space System Search of Vehicles, 108
space station, 109
Spain, 9, 21
special inds., land mob. rad. svc, 52, 108
special marine warning, 110
spectrometers, 15
spectrum, radio, 88
spin-scan radiometer, 19
Spot, satellite, 23, 24
spread spectrum, 109
ssb, 55, 87, 109
SSFC, 12
sstv, 55, 109
St. Johns, 42
standard time stations, 48
    frequencies, 49
    weather, 48
station, 110
stationary orbit, satellite, 18, 23, 28, 17
statute mile, 110
storm, name, 37
    plots, 20
    surge, hurricane, 38
    track, hurricane, 37
    tracking nets, 54
    warning, 110
stratosphere, 6
stratospheric sounder, 16
submillimeter, 110
Sudan, 11
Suitland, Maryland, 8, 19
Sun, 10
    angle, 17
    magnetic storms, 44
Sun-synchronous orbit, 16
Sunflower, Japan, satellite, 22
Sunspot Cycle, 44
sunspots, 44
sunstroke, 110
super high frequency, 109
suspended water droplets, 7
Sweden, 21, 25
Switzerland, 21
SWL amateur radio nets, 86, 87
swl, 110
Sydney VOLMET, 41
Synchronous Meteorological Sats, 18
Syncom, 18
synoptic meteorology, 6
    weather maps, 5

TTT

Taipei VOLMET, 41
Taiyuan Space Center, 22
Tananarive VOLMET, 42
Tanegashima Island launch, 22
technology, latest, 20
Tehran VOLMET, 42
Tel Aviv VOLMET, 41
telegraph, 5
telemetry, 110
    RECCO code, 39
teleprinter, 108, 109
telescope, infrared, 19
    visible-light, 19
television, 88, 110
temperature, 10, 110

terminal node controller, 31, 33
Texas, 87
	Corpus Christi, 47
	Galveston, 47
thermodynamics, 6
thermometer, 5
third-generation satellite, 16
three-dimensional photos, 19
thunderstorm, 8
	severe, 110
time stations, 48
	frequencies, 49
	satellite, 51
	weather, 48
	worldwide, 49
time-lapse photos, 14
Tirat Tsvi, Israel, 9
TIROS Operational System, 15
TIROS satellites, 14, 16, 17, 18
TIS, 52, 53
	NOAA Weather Radio, 53
tnc, 31, 33
Tokyo, 22
Tokyo VOLMET, 41
TOR, 48
tornado, 8
	warning, 7, 110
	watch, 110
TOS, 15
Toulouse VOLMET, 41
Tours VOLMET, 41
tracking progams, satellite, 29
trade winds, 10
traffic, 110
	emergency, 54
transfer orbit, satellite, 17
Traveler's Information Service, 52, 53
trawlers, MBs, 51
Trenton, 42
Tripoli VOLMET, 42
tropical, 10
tropical, ocean project, 8
tropical storm, 14
	warning, 110
troposphere, 6
troughs, upper-level, 14
Tuguegarao, Philippines, 9
Tully, Queensland, 11
Turk, 27
Turkey, 21
turnstile antenna, 28
tv, 110
	fast-scan, 54
	slow-scan, 54
TVRO, WEFAX, 30
two meters, 110
typhoon, 8, 14, 110

UUU

US Coast Guard, 27, 42, 47, 108, 109, 110
US Department of Transportation, 108
US Forest Service, 24
US Weather Bureau, 6
uhf, 22, 25, 55, 87, 88, 110
	amateur radio, 54
	propagation, 45
	receiver, 27
ulf, 55, 110
ultra high frequency, 55, 110
ultra low frequency, 55, 110
ultraviolet light, 88
United Nations, UN, 8
United States Air Force, 110
United States Navy, 110

United States, 110
universal time, 7
University of Alabama, 16
updates, frequencies, 87
uplink, 110
upper sideband, 55, 110
USAF, 109, 110
usb, 41, 55, 87, 110
USCG, 109, 110
USN, 109, 110
USSR, 23, 25
	launches, 23
	weather satellites, 15, 21
		first weather satellite, 15
Ust'Shchugor, Russia, 9
UTC, 7, 56, 110

VVV

Valdez, Alaska, 24
Vancouver VOLMET, 41
Vandenberg AFB, California, 17, 20
VAS, 19
vertical atmospheric sounder, 19
very high frequency, 55, 110
very important person, 110
very low frequency, 55, 110
vhf, 22, 25, 27, 53, 55, 87, 88, 110
	amateur radio, 54
	high-band, 52
	propagation, 45
	receiver, radio, 27
video, 30
Vienna VOLMET, 41
VIP, 110
Virginia, Alexandria, 48
	Norfolk, 47
	Portsmouth, 48, 109
	Wallops Island, 19
visibility, 10
visible light, 88
	photos, 14
	telescope, 19
vlf, 55, 110
Voice of the Natl Weather Svc, 13
voice message codes, VORTEX, 38
volcanic eruptions, 20
VOLMET, 41
	frequencies, 41
	aviation weather, 41
volunteer emergency group, 54, 56
VORTEX voice message codes, 38
Vostok, 9

WWW

W1AW, ARRL station, 87
	emergency bulletins, 86
Wadi Halfa, Sudan, 11
WAE, public coast station, 47
WAK, public coast station, 47
Wake VOLMET, 41
walkie-talkie, 108
Wallops Island, Virginia, 19
warnings, 110
	NAVAREA, 45
	severe weather, 7
	tornado, 7
Warsaw VOLMET, 41
Washington, Seattle, 47
watch, 110
water vapor, 20
water, suspended droplets, 7
wbfm, 108, 110
WDR, public coast station, 47

weather, 56, 110
	aviation, 41
	crises, amateur radio, 54
	facsimile, 22
	forecasting, 5
		radio, 6
	frequencies, 87, 88
	fronts, 14
	graphs, 7
	maps, 27
	patterns, 8
	plots, 7
	predicting, 8
	tracking nets, 54
	weather band receivers, 13
	Weather Bureau, 6, 14
	Weather Radio stations, NOAA, 13
weather satellites, 7, 12, 14, 30
	frequencies, 22
	receiver, 29
	how to receive, 27
WEFAX, 22
	gear, 31
	how to receive, 27
	radio, 27
		antenna, 28
	shortwave, 33
		frequencies, 33
		receive, 33
	transmission schedule, 35
	TVRO, 30
West Africa region VOLMET, 42
Wexler, Harry, 14
WFA, public coast station, 47
WGB, public coast station, 47
wide band frequency modulation, 110
wind, 10
	hurricane, 38
	speed, 110
	upper-level, 14
Wind and Cloud, satellite, 22
wind chill index, 110
wind profilers, 7, 87
wind-sensing radar, 7
Wisconsin, Milwaukee, 10
WLO, public coast station, 47
WMH, public coast station, 47
WNJ, public coast station, 47
WOM, AT&T station, 7, 8, 46, 51
WOO, AT&T station, 46
World Meteorological Org., 7, 8, 51
World War II, 8
World Weather Watch, 8
worldwide time stations, 49
WOU, public coast station, 47
WP-3D Orion aircraft, 37
WQX, public coast station, 47
WWV, time station, 8, 48
WWVH, time station, 8, 48

XXX

X-band, 88, 110
X-rays, 23, 88

YYY

Yellowstone National Park, 24
Yesilkoy VOLMET, 42
Yugoslavia, 11
Yukon, 9

ZZZ

Z, zulu time, 7, 110

DALLAS REMOTE IMAGING GROUP

Satellite Imagery Bulletin Board

(214) 394-7438

(300 - 14,400 BPS   V.32bis   V.42)

Weather Satellite Images from 22,300 miles in Space
Thousands of NASA *Voyager Spacecraft* Images
Image Processing Programs for Editing
***Satellite Tracking* Programs - USAF Keplerians**
Digital Signal Processing Programs of Telemetry
Hundreds of Satellite Frequencies
APT, WEFAX, HRPT, VAS *Weather Satellite* Images
Large Satellite database for Orbital Analysis
INTERNET ACCESS FOR E-MAIL AND NEWS GROUPS
WORLDWIDE Satellite Tracking *Intelligence Network*

6 TELEPHONE LINES WITH USR DUAL STANDARD MODEMS FOR EASY ACCESS TO THE NETWORK!- NO WAITING FOR BUSY LINES!!

TO JOIN THE DALLAS REMOTE IMAGING GROUP BBS:
  * USE *VISA* OR *M/C* ON-LINE REGISTRATION
  * SEND A CHECK FOR $ 32.50 (TAX INCLUDED) TO:

DALLAS REMOTE IMAGING GROUP
P. O. BOX 117088
CARROLLTON, TEXAS 75011-7088

(FAX - 214-492-7747)

Monitoring NASA Communications

How to tune in the National Aeronautics & Space Administration on Shortwave, VHF, UHF and Satellites

by Anthony R. "Tony" Curtis K3RXK

$14.95 + $2 Shipping & Handling   Visa/MasterCard
or see your Radio Hobby Book Dealer

TIARE PUBLICATIONS
P.O. Box • Lake Geneva, WI 53147

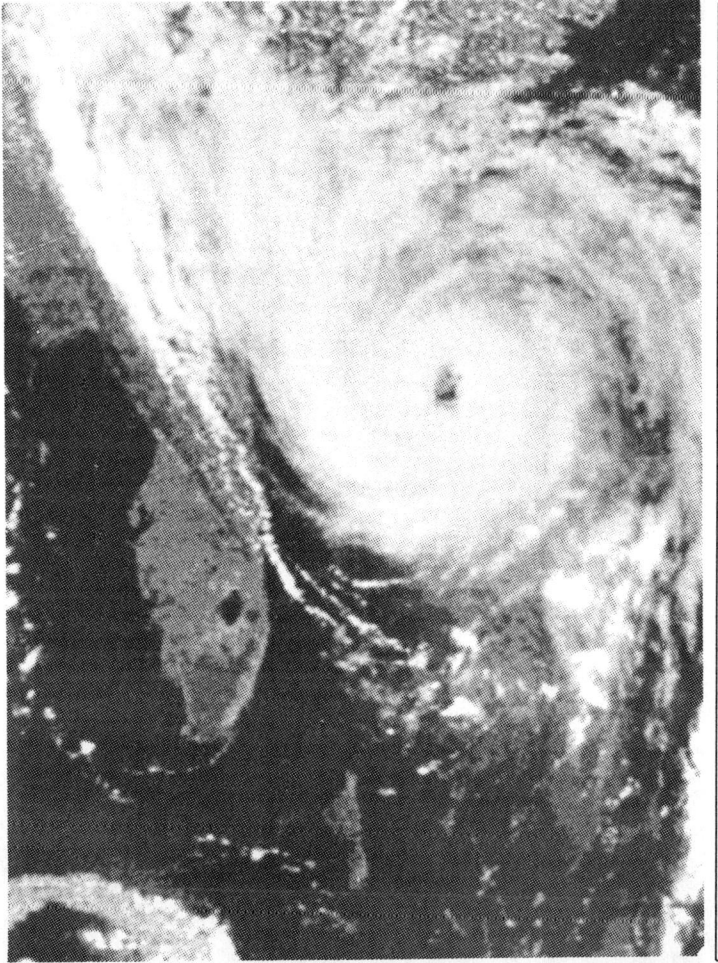

RECEIVE PICTURES LIKE THIS ON YOUR PC !

- IBM and IBM Compatible PCs (512k Memory) -
- Straight Edges (PLL Sampling) -
- NOAA, Meteor, GOES, HF Fax -
- 256 Colors/64 Grays with SVGA -
- Zoom, Color, and Gray Control -
- Dot Matrix and Laser Printer Support -
- 4800 8-Bit Samples per Second -
- Full 12 Minute Pass with NOAA -
- Latitude/Longitude/Map Overlay Software -
- Fully Assembled and Tested -

Requires: 8 or 16-Bit Slot, VGA,
Hard Disk or RAM Disk (4MB free space),
Receiver, and Simple Antenna (Dish not Required)

MultiFAX Card, Software, and Full Instructions
Introductory Price $249 Including S&H (in US)

Demo Kits $15 - Software Upgrades $20
MasterCard/VISA

For more information and orders please write to:

MultiFAX
143 Rollin Irish Road - Dept WR
Milton, VT 05468

The WeatherBASE Microcomputer Products

from
L'Softworks Limited

The WeatherBASE programs consist of a growing set of microcomputer applications that assist the weather observer in recording, viewing, modifying, and analyzing weather data.

Products & Features

LWEATHER. History & User Data Log: $50
Day almanac with station all-time history. Also includes user station log keeper.

DAILY. Database Manager: $50
A daily station record and climatological analysis application with your data or NCDC's.

ALMANAC. Weather & Astronomical: $25
A daily planet & weather almanac.

HISTORY. Database Manager: $25
Station all-time history your data or NCDC's.

MAPPER Weather Data Mapping: $25
Data can be downloaded or manually entered.

LOGGER Database Manager: $25
User station log keeper & reporter.

FORECASTER. Day Forecaster: $25
Weather forecasting at your computer.

Product Ordering

Write a check or money order payable to:
L'Softworks Limited
803 - 12th Avenue NW : New Brighton, MN 55112-2666

Demo Disk $5.00 Specify 3½" or 5¼" DD or HD
(IBM-PC format 360kb, 1.2mb, 720kb or 1.44mb)
Phone Information: (612) 636-5538

M1000 = RTTY/FAX Excitement

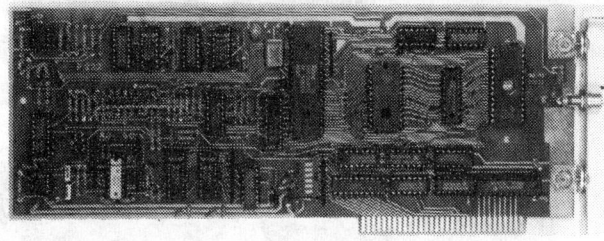

Turn your IBM computer (or clone) into a powerful intercept device! The **Universal M-1000** Decoder Card requires just one slot in your PC. Your computer can open up a new world of listening (and *seeing*) opportunities! You can decode standard modes such as Morse Code, Baudot RTTY and Sitor A/B. Advanced diplo.-military modes such as ARQ-M2, ARQ-E and ARQ-E3 are also supported. ASCII and Packet modes are even featured. For FAX reception (only) your computer must have either an EGA or VGA monitor (color or mono). The video quality of your FAX intercepts will amaze you. Advanced FAX imaging includes false-color and zoom features. FAX images as well as text traffic can be saved to disk for later retrieval or analysis. Despite the sophistication of this device, operation is easy through on-screen menus, status indicators and help windows. A new *datascope* feature operates in both RTTY and FAX modes. The M-1000 comes with an informative manual and software on both 3½" and 5¼" diskettes. **Only $399.95 (+$5).**

UNIVERSAL RADIO
6830 Americana Pkwy.
Reynoldsburg, OH 43068
☎ Orders: 800 431-3939
☎ Info.:   614 866-4267

⇨ Call or write to receive your **FREE** 100 page catalog.

Universal has been serving radio enthusiasts since 1942.

WEATHERTRAC

Weather Instrumentation For Business, Home, Industry & Schools

- Complete Weather Stations
- Computer Compatible Systems
  - Barometers
  - Rain Gauges
  - Thermometers
  - Hygrometers
  - Wind Sensors
  - Lightning Detectors

FREE CATALOG   DISCOUNT PRICES
WRITE OR CALL

WEATHERTRAC
P.O. Box 122
Cedar Falls, Iowa 50613
319-266-7403   800-798-8724
VISA AND MASTERCARD

Computer Aided Scanning
a new dimension in communications from Datametrics

Now you can enhance your ICOM communications receiver through a powerful computer controlled system by Datametrics, the leader in Computer Aided Scanning. The system is as significant as the digital scanner was five years ago and is changing the way people think about radio communications.

- The Datametrics Communications Manager provides computer control over the ICOM R9000, R7000 or R71A receiver.
- Powerful menu driven software includes full monitoring display, digital spectrum analyzer and system editor.
- Innovative hardware design requires no internal connections.
- Comprehensive manual includes step by step instructions, screen displays, and reference information.
- Extends ICOM capabilities including autolog recording facilities, 1000 channel capacity per file, and much more.
- Overcomes ICOM limitations such as ineffective scan delay.

Datametrics, Inc

___ R7000 or R71A system $ 349
___ R9000 system $ 1,599
___ Manual and demo disk $15

Requires ICOM receiver and IBM PC with 640K and serial port. The R71A version also requires an ICOM UX-14.

Send check or money order to Datametrics, Inc., 2575 South Bayshore Dr, Suite 8A, Coconut Grove, Fl, 33133. 30 day return privileges apply.

TIMESTEP

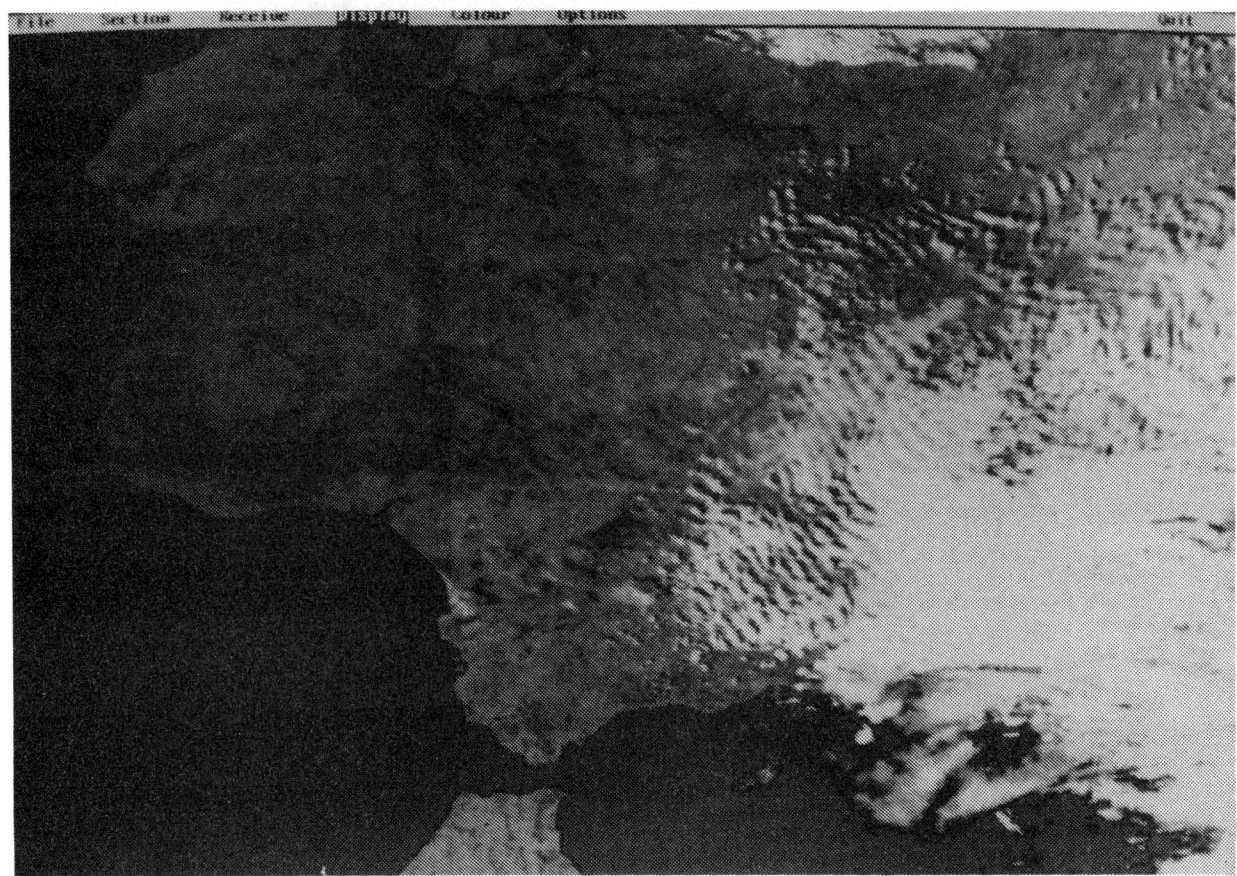

HRPT System

Noise-free digital HRPT transmissions from NOAA, with a ground resolution of just 1.1km, allow images to be received in incredible clarity. Rivers, lakes, mountains, cities and even small towns can be seen on good days. Fishermen will appreciate the increased resolution of sea surface temperatures.

Image processing, including variable and histogram contrast equalisation combined with full colour editing, gives the best possible results from any image. Colour enhancement allows sea surface temperature and land details to stand out in high contrast. Any number of colour palettes can be saved for future use. The sophisticated mouse-driven software allows all five bands to be saved and displayed on nearly all VGA and SVGA cards right up to 1024 pixels, 768 lines and 256 colours.

Zoom to greater than pixel level is available from both a mouse-driven zoom box or using a roaming zoom that allows real time dynamic panning.

Sections of the image may be saved and converted to GIF images for easy exchange.

Latitude and longitude gridding combined with a mouse pointer readout of temperature will be available late in 1991.

Tracking the satellite is easy and fun! Manual tracking is very simple as the pass is about 15 minutes long. A tracking system is under development and expected by the end of 1991. A 4-foot dish and good pre-amplifier are recommended. The Timestep Receiver is self-contained in an external case and features multi-channel operation and a moving-coil S meter for precise signal strength measurement and tracking. The data card is a Timestep design made under licence from John DuBois and Ed Murashie.

Complete systems are available, call or write for a colour brochure.

USA Amateur Dealer. Spectrum International, P.O. Box 1084, Concord, Massachusetts 01742. Tel: 508 263 2145

TIMESTEP WEATHER SYSTEMS
Wickhambrook Newmarket CB8 8QA England Tel: (0440) 820040 Fax: (0440) 820281

VGASAT IV & MegaNOAA APT Systems
1024 x 768 x 256 Resolution and 3D

The Timestep Satellite System can receive images from Meteosat, GOES, GMS, NOAA, Meteor, Okean and Feng Yun. Using an IBM PC-compatible computer enables the display of up to 1024 pixels, 768 lines and 256 simultaneous colours or grey shades depending on the graphic card fitted. We actively support nearly all known VGA and SVGA cards. Extensive image processing includes realistic 3D projection.

100 Frame Automatic Animation

Animation of up to 100 full screen frames from GOES and Meteosat is built in. We call this 'stand alone animation' as it automatically receives images, stores them and continuously displays them. Old images are automatically deleted and updated with new images. The smooth animated images are completely flicker-free. Once set in operation with a single mouse click, the program will always show the latest animation sequence without any further operator action.

NOAA Gridding and Temperature Calibration

The innovative MegaNOAA program will take the whole pass of an orbiting satellite and store the complete data. Automatic gridding and a 'you are here' function help image-interpretation on cloudy winter days. Spectacular colour is built in for sunny summer days. Self-calibrating temperature readout enables the mouse pointer to show longitude, latitude and temperature simultaneously.

Equipment

Meteosat/Goes
- ❏ 1.0M dish antenna (UK only) ❏ Yagi antenna
- ❏ Preamplifier ❏ 20M microwave cable
- ❏ Meteosat/GOES receiver
- ❏ VGASAT IV capture card
- ❏ Capture card/receiver cable
- ❏ Dish feed (coffee tin type)

Polar/NOAA
- ❏ Crossed dipole antenna
- ❏ Quadrifilar Helix antenna (late 1991) ❏ Preamplifier
- ❏ 2 channel NOAA receiver ❏ PROscan receiver
- ❏ Capture card/receiver cable

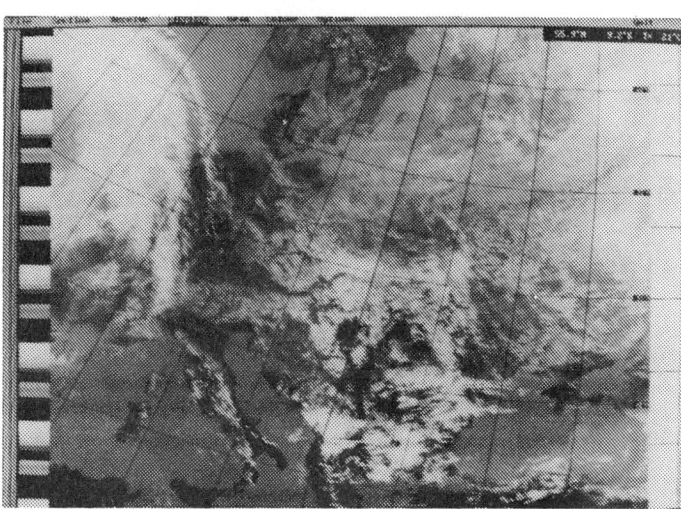

Call or write for further information.

USA Education Dealer. Fisher Scientific, Educational Materials Division, 4901 W. LeMoyne Street, Chicago, IL 60651.
Tel: 1-800-621-4769

USA Amateur Dealer. Spectrum International, P.O. Box 1084, Concord, Massachusetts 01742. Tel: 508 263 2145

TIMESTEP WEATHER SYSTEMS
Wickhambrook   Newmarket   CB8 8QA   England   Tel: (0440) 820040   Fax: (0440) 820281

What you see is what you get - No Kidding!

Actual image printed using PC Paintbrush IV Plus on an HP Laserjet IIIP

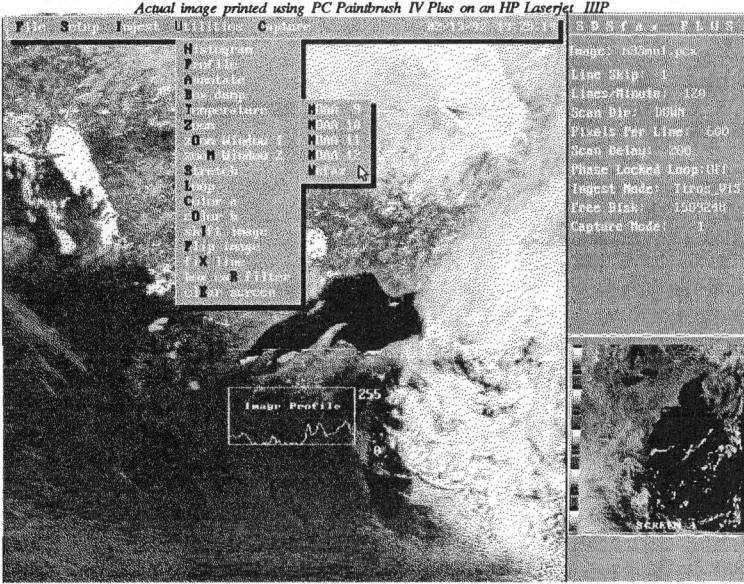

SDSfax PLUS

- o Super VGA (800x600)
- o 256 gray level imaging
- o Histogram / Stretch
- o Profile image data
- o Zoom, Pan, Scroll
- o PCX, Binary, Ascii files
- o Color/Gray, Enhancement
- o Precise temp calibration using telemetry wedges
- o Tag, Loop, Sort files
- o Auto-acquire modes
- o Graphic signal level meter
- o Box dump of pixels
- o Annotate images, 3 fonts
- o Full editing of pixel colors
- o Many more features....

Professional Quality Weather Satellite Images and Processing on your PC

If your a discriminating Weather Satellite user, who needs a company that can backup what it sells - You have found SATELLITE DATA SYSTEMS, INC.

We offer an IBM/PC card (PC through 486's), Software and complete hardware systems for APT, WEFAX and GOES-TAP. (Dishes, antennas, preamps, receivers and other hardware)

CALL OR WRITE FOR OUR CATALOG AND FULL PRICING SHEETS
FREE DEMO's, VGA and/or SVGA - Please specify 5-1/4" or 3-1/2"

SDSfax standard VGA software:

Supports 640 x 480 x 16

Integrated Satellite Tracking & Scheduling

Zoom, Label, Loop up to 99 images

Temperature Measurment

Timer functions with Auto Acquire Programmable for 24 hr or greater periods

Print to most laser & dot matrix printers

Enhanced images using look-up-tables (LUTs), 'Equivalent" to 64 gray shades

APT - WEFAX - GOES-TAP - MAPS

Full Palette control

OPTION: Latitude,Longitude to pixel x,y

Our WeatherFax cards have been laboratory tested and verified to comply with FCC regulations under part 15.

Our products are **MADE IN THE USA**

We offer Educational Discounts and site licensing.

Our business is dedicated, full time, to low-cost weather satellite receiving systems. We are celebrating 29 years in business, and have been in weather exclusively for 8 years. We will be around to support you.

PRICING:
Card with SDSfax $ 799.00
Card with SDSfax PLUS $ 899.00
SDSfax PLUS only $ 195.00

SATELLITE DATA SYSTEMS, INC.
P. O. Box 219
Cleveland, MN 56017-0219
(507) 931-4849